Mechanical Characteristics of the Prestressed Reinforced Concrete Lining with Unbonded Annular Anchors

CAO Ruilang WANG Yujie PI Jin

·北京·

图书在版编目（CIP）数据

无黏结预应力环锚衬砌传力机制和特性 = Mechanical characteristics of the prestressed reinforced concrete lining with unbonded annular anchors : 英文 / 曹瑞琅，王玉杰，皮进著. -- 北京 : 中国水利水电出版社，2022.12
ISBN 978-7-5226-1471-7

Ⅰ. ①无… Ⅱ. ①曹… ②王… ③皮… Ⅲ. ①预应力锚夹具－衬砌－研究－英文 Ⅳ. ①TV672

中国国家版本馆CIP数据核字(2023)第057595号

书　　名	**Mechanical characteristics of the prestressed reinforced concrete lining with unbonded annular anchors**
中文书名	**无黏结预应力环锚衬砌传力机制和特性** WUNIANJIE YUYINGLI HUANMAO CHENQI CHUANLI JIZHI HE TEXING
作　　者	CAO Ruilang（曹瑞琅） WANG Yujie（王玉杰） PI Jin（皮进）
出版发行	中国水利水电出版社 （北京市海淀区玉渊潭南路1号D座　100038） 网址：www. waterpub. com. cn E - mail：sales@ mwr. gov. cn 电话：（010）68545888（营销中心）
经　　售	北京科水图书销售有限公司 电话：（010）68545874、63202643 全国各地新华书店和相关出版物销售网点
排　　版	中国水利水电出版社微机排版中心
印　　刷	天津嘉恒印务有限公司
规　　格	170mm×240mm　16开本　9印张　174千字
版　　次	2022年12月第1版　2022年12月第1次印刷
定　　价	**80.00**元

Preface

For large-section pressure tunnels in hydropower projects, municipal projects and long-distance water diversion projects, normal reinforced-concrete linings are prone to cracking under the internal water pressure. Large cracks in a lining not only cause the loss of water heads but also affect the stability of the tunnels. Although steel fiber-reinforced concrete can improve the low tensile strength of the concrete, it is still difficult to cope with the tensile stress of the lining caused by high internal water pressure. Steel lining requires high-quality structural welding, and since the tunnel lining on the contact side of the surrounding rock cannot be welded, the single-sided welded steel is prone to erosion and buckling under high internal and external water pressures.

With the rapid development of prestress technology, the design concept of "reinforcing tensile strength via high compressive strength of concrete" is gradually being introduced into the design of the support structure of pressure tunnels to prevent the formation of lining cracks. High-pressure grouting into the seam between the lining and the surrounding rock could not only reinforce the rock body but also generate prestress in the anchor lining. At present, the grouting of prestressed tunnel linings has been successfully applied in deep and long pressure tunnels. However, high-pressure grouting requires that the surrounding rock must provide all the counterforces for the prestress, and this method can hardly be applied to a shallow tunnel with weak surrounding rock strength; otherwise, it may cause rock cleavage.

To reduce the pressure of the surrounding rock and make the lining resist high internal water pressure by relying only on its own prestressing, a high-strength anchor was applied in the pressure tunnel to form an active prestressed anchor. In the US, the underground water pipeline network adopted a prestressed concrete cylinder pipe (PCCP), but the prefabricated PCCP can only be used in an open tunnel excavation on a road instead of in a subsurface excavation. In Japan, the technology of assembling prestressed segments in the construction of underground drainage projects was put forward, but the prestressed segments are generally applied only in shield tunnels. To achieve the field pouring of a prestressed lining, the technology of a prestressed anchor lining was proposed and applied in the surge tank in the Crimsel-Oberrar water pumped storage power station in Switzerland; however, it is a structure of single annular anchor that can provide limited prestress with which it is difficult to meet the needs of large cross section tunnels coping with high internal water pressure. Thus, in the construction of the desilting tunnel in the China Xiaolangdi multipurpose project and the Dahuofang water delivery tunnel, a new prestressed reinforced concrete lining with multi-layer (up to 2) multi-hoop (up to 2) unbonded annular anchors (hereinafter referred to as MUAA lining) has been developed. Furthermore, the new type of prestressed lining with a single-layer double-hoop unbonded annular anchor was proposed through construction of the long-distance water delivery tunnel for the water transfer project from the Songhua River (built in 2014-2020). Engineering practices show that prestressed lining with annular anchors has advantages of low prestress loss and uniformly distributed inward pressure, while, it also shows complex stress and strain mechanism between the annular anchor and the lining concrete.

To ensure the long-term operational safety of the MUAA lining and promote the popularization and application of this technology, we adopt it in the practical application of our project, based on the existing data investigation and analysis, and comprehensively use a variety of approaches, such as theoretical analysis, numerical simulation, indoor tests, and field tests. The mechanical characteristics, structural form, construction method, and quality control are extensively studied. This book systematically summarizes the latest achievements of the MUAA lining, which is expected to play a vital role in promoting the theoretical development and engineering practice of pressure tunnel support structure design.

In the process of compiling this book, we received the support and help of ZHAO Yufei, QI Wenbiao, JIANG Long, XUE Xingzu, ZHENG Lifeng, SUN Xingsong, LIU Yang, WANG Qian, and other comrades. The authors would like to express their sincere thanks to them.

Authors

May, 2022

Contents

Chapter 1
Structural characteristics and design of the MUAA lining

1.1 Introduction

With the rapid development in the engineering of underground construction worldwide, large-diameter tunnels with high internal water pressures have emerged in many important hydropower, municipal projects and long-distance water diversion projects in recent years, such as the tunnels of South-North Water Transfer Project through to the Yellow River and Songhua Water Transfer Project from Songhua River to Jilin Province in China (Shen and Liu, 2003; Pi et al, 2018), the tunnel of the construction works for underground rainwater storage in Osaka, Japan (Nagamoto et al, 2008), the diversion tunnel of Tarbela Project in Pakistan (Tate and Farquharson, 2000), and the pressure tunnel at the Salween Power Station in Myanmar (Kirchherr and Julian, 2017). The design and construction of appropriate supports which resist the internal and external pressures exerted by internal water and the surrounding rock mass for pressure tunnels is much more difficult than that for non-pressure tunnels. It is a huge challenge to select and design a suitable lining or support structure, especially when a tunnel experiences problems like large diameter, poor surrounding rock quality, thin overburden, and high internal pressure (Simanjuntak et al, 2014).

To cope with this technical challenge, concrete linings actively prestressed by posttensioning steel are preliminarily proposed. The prestress is induced by a prestressing tendon (or an annular anchor) which runs in or around the concrete ring. Such actively prestressed linings have been used in the construction of the Crimsel Tailrace tunnel surge shaft in Switzerland (Trucco and Zeltner, 1978) and in the Italian Presenzano pressure tunnel (Matt et al, 1978). In the early stage, the prestressed concrete lining exhibited a simple structure and the active prestress was provided only by a single hoop of anchor, which could not fulfil the needs of large-section tunnels under high internal water pressures. Since the introduction of this type of lining structure to China in the late 20th century, it has been developed into a new prestressed reinforced concrete lining with multi-layer (up to 2) multi-hoop (up to 2) unbonded annular anchors (hereinafter referred to as MUAA lining) in China (Cao et al, 2016; Cao et al, 2019; Cao et al, 2021; Wang et al, 2020), as shown in Fig. 1.1.

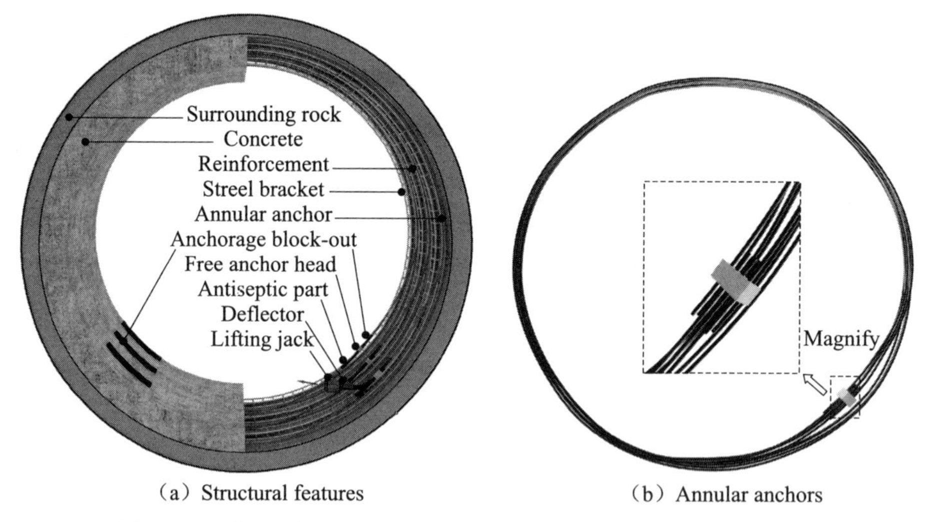

(a) Structural features (b) Annular anchors

Fig. 1.1 3-D effect diagram of the new-typed MUAA lining

In the MUAA lining, the concrete creates radial precompression stress through tensioning the unbonded annular anchors by the jack, which makes full use of the superior tensile and anti-seepage function of the prestressed concrete structure. It can reduce the load sharing of surrounding rock by increasing the active prestress of anchors and is not restricted by the geological conditions of surrounding rock. The engineering applications show that the MUAA lining's advantage lies in uniform stress on the whole lining, greatly reduced prestress loss, less steel strand material, and smaller prestressed weak area.

1.2 Structural systems

Both ends of the annular anchors are anchored on the same free anchor head in the reserved anchorage block-out, and the annular anchor extrudes the reinforced concrete through the angle deflector and the tensioning of the jack, so that the lining can produce the annular prestress. Then, the micro-expansion concrete helps the anchorage block-outs to close, so that a uniform prestressed structure is formed in the lining. Therefore, the unbonded annular anchor and the anchoring system are the source of the prestressing of the lining, the tension is the energy source, the reinforced concrete of the lining is the carrier to realize the prestressing of the structure, and the anticorrosion system is the long-term and effective guarantee of the MUAA lining.

In view of this, the structural systems of the MUAA lining can be divided into anchoring system, tensioning system, reinforced concrete system, anticorrosion system and waterproof structure. The structural systems composition of the MUAA lining is shown in Fig. 1.2.

1.2.1 Anchoring system

The anchoring system is mainly composed of unbonded annular anchor, free anchor head, operation clip, rubber splint, steel splint and screw. The schematic diagram of unbonded prestressed annular anchor principle is shown in Fig. 1.3.

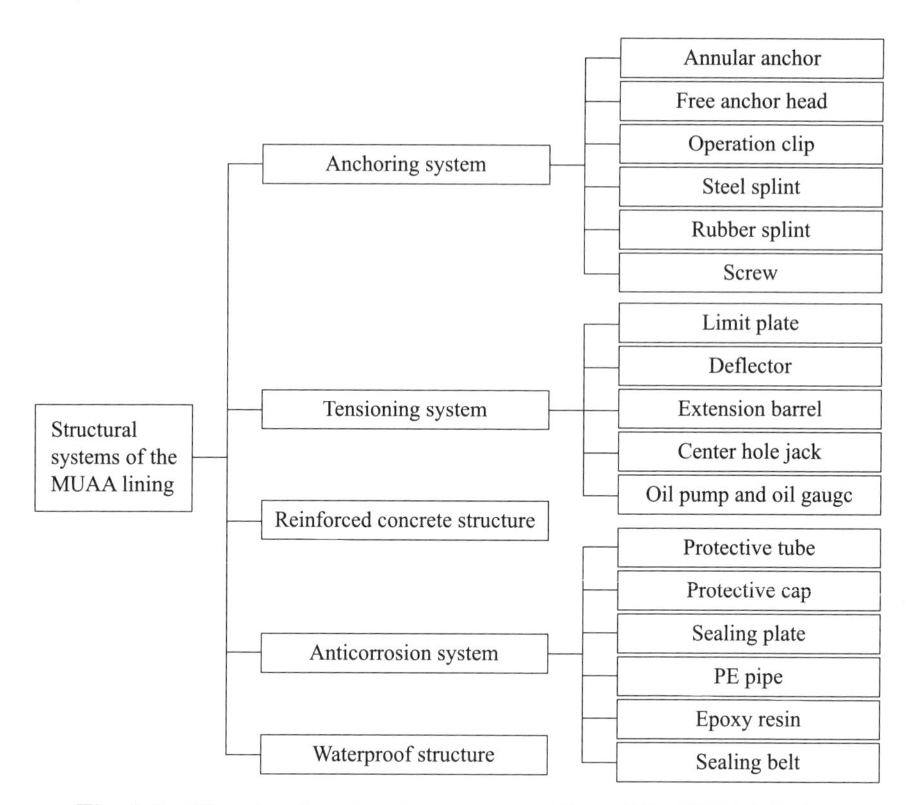

Fig. 1.2　The structural systems composition of the MUAA lining

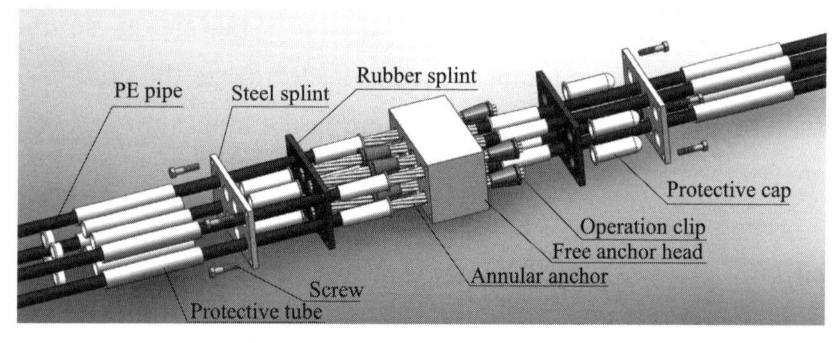

Fig. 1.3　The schematic diagram of the anchoring system

The free anchor head is the core force unit of the tensioned anchoring in the entire annular prestressed system. In actual construction, a free anchor head with anchoring end and tensioning end is adopted, where the anchor hole of the tensioning end is designed in the middle. The anchor hole of the anchoring end is set on both sides of the free anchor head, which realizes a compact appearance of the anchor plate and uniform prestressing distribution. The annular anchor operation clip is a key part of the anchoring system, which requires high quality alloy steel, and the processing and heating treatment must be strictly controlled. There are taper holes on the free anchor head, which cooperate with the clip, and the anchors is anchored by the wedge tightening principle of taper holes. The free anchor head and the operation clip of the unbonded prestressed anchoring system are subjected to long-term load, so it is particularly critical to ensure the effectiveness of the anchoring.

Unbonded prestressed annular anchor generally uses 7ϕ 5 mm high strength and low relaxation steel strand, which should meet the China Standard GB/T 5244-2014 *Steel Strand for Prestressed Concrete* and the American Standard ASTM A416/A416M-2010 *Standard Specification for Steel Strand, Uncoated Seven-Wire for Prestressed Concrete*. The main parameters are shown in Table 1.1.

Table 1.1 Physical and mechanical parameters of prestressed steel strand

Parameter	Diameter	Standard strength	Sectional area	Failure load	Elastic modulus	Design strength
Symbol unit	D / mm	f_{ptk} / MPa	A_p / mm^2	f_{ptk} / kN	E_s / GPa	f_p / kN
Value	15.24	1860	140	260.5	195	195.4

In addition, according to the Chinese standard, the thickness of PE pipe should not be less than 1.5 mm, the deviation coefficient k is not more than 0.04, and the friction coefficient μ between steel strand and PE pipe is not more than 0.1.

1.2.2 Tensioning system

Reasonable, quick and accurate annular anchor tensioning is an important guarantee for the construction quality of prestressed annular anchor, and the selection of annular anchor tensioning equipment is also significant. The annular anchor tensioning system is mainly composed of limit plate, deflector, extension barrel, center hole jack, oil pump and oil gauge [Fig. 1.4(a)]. Fig. 1.4(b) is a working drawing of the tensioning system on site.

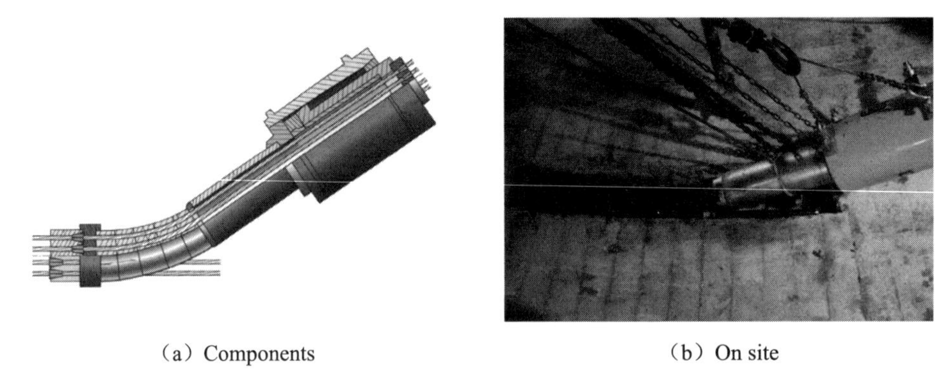

(a) Components (b) On site

Fig. 1.4 Working drawing of unbonded prestressed annular anchor tensioning system

The annular anchor can only be tensioned at varying angles by the deflector. The friction loss of prestress is closely related to the deflector. The deflector used in the field test is a steel structure optimized from multi-section combination to two-section bend type, which is not only easy to disassemble and assemble, but also greatly reduces the friction loss rate.

The center core jack is the executive element used with the prestressed anchor, and the oil pump is the power source adapted to the type of the jack. The on-site tensioning should be based on the principle of "light weight and easy construction". YCW jack and ZB4-500 oil pump are usually used (Fig. 1.5). Equipment parameters are shown in Table 1.2 and Table 1.3. In engineering application, jack and oil gauge should be calibrated before annular anchor lining is tensioned.

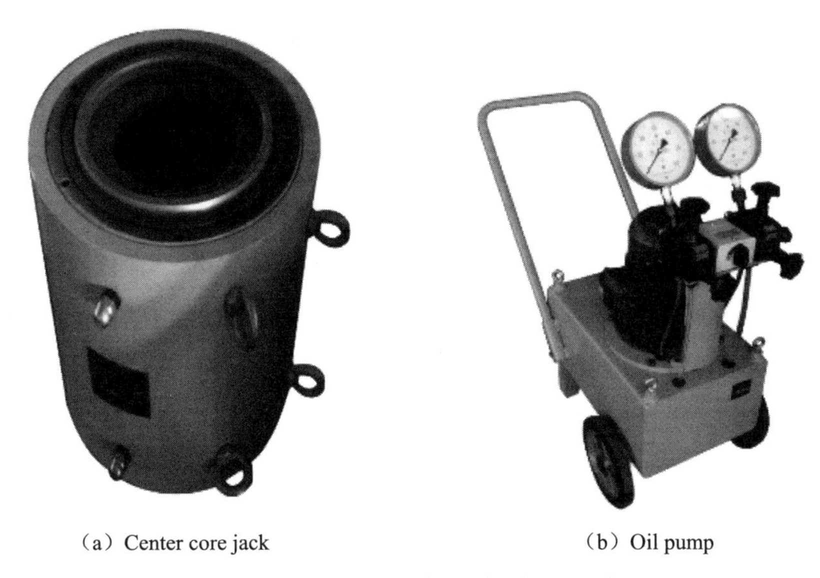

(a) Center core jack (b) Oil pump

Fig. 1.5 Photograph of tensioning equipment

Table 1.2 Parameters of YCW series jack

Type	Tensile load / kN	Oil pressure / MPa	Return pressure / MPa	Center hole diameter / mm	Range / mm	Weight / kg	Appearance size / mm
YCW100B	973	51	$<$ 25	ϕ78	200	65	370×ϕ214
YCW150B	1492	50	$<$ 25	ϕ120	200	108	370×ϕ285
YCW200B	1998	53	$<$ 25	ϕ120	200	135	382×ϕ310
YCW250B	2480	54	$<$ 25	ϕ140	200	164	380×ϕ344

Table 1.3 Parameters of oil pump

Type	Rated pressure / MPa	Standard flow value / (L/min)	Weight / kg	Appearance size / mm
ZB4-500	50	2×2	120	745×494×1025

1.2.3 Anticorrosion system

The anticorrosion system is the long-term and effective guarantee of the MUAA lining (Fig. 1.6). HDPE (High Density Polyethylene) anticorrosion casing is made of a polymer material, which is waterproof, moisture-proof and corrosion resistant.

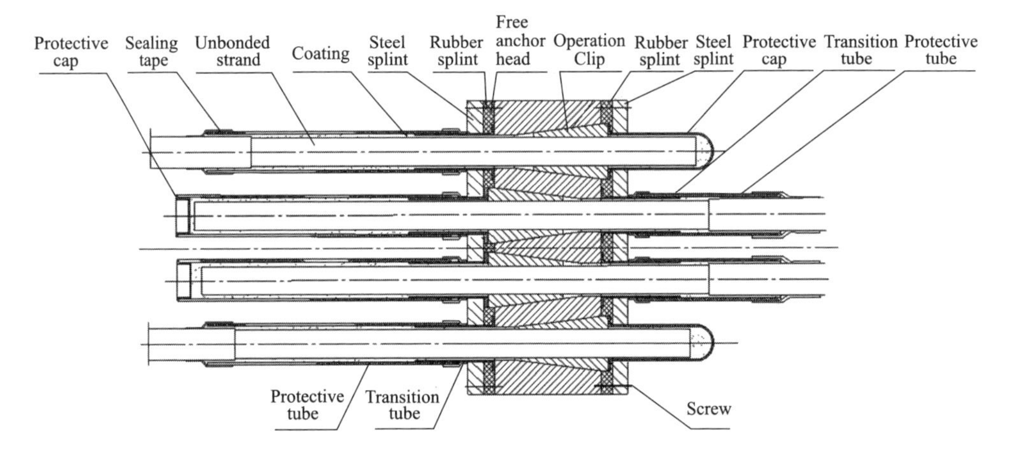

Fig. 1.6 Schematic diagram of anticorrosion system

1.2.4 Reinforced concrete structure

Reinforced concrete in the MUAA lining can be consistent with conventional reinforced concrete. For example, a project uses Q235 steel with a 22 mm diameter threaded bar for the main bar and an 18 mm diameter threaded bar for the stirrups. The concrete is made of ordinary Portland cement, the strength grade is C40, the design value of axial compressive strength is 19.1 N/mm², and the elastic modulus is 3.25×10^4 MPa. The aggregate is mixed in two levels, and the concrete mix ratio design is shown in Table 1.4.

Coarse and fine aggregate are both dry on saturated surface. Coarse aggregate is secondary graded crushed rock (medium stone: small stone = 60:40). The admixture is YL-5 high performance water reducing agent (liquid), and the amount of admixture is 1.0% of the dosage of cementing materials. The air content of concrete is 4.6%.

Table 1.4 Mix proportion design of concrete in the MUAA lining

Design standard	Cement type	Sand ratio /%	Slump / mm	Water-cement ratio
C40F150W12	P.O 42.5	42	179	0.32

1.2.5 Waterproof structure

The construction joints enable the MUAA linings have independent mechanical properties (Fig. 1.7). In order to ensure the longitudinal sealing of the lining, rubber waterproof belt (specification: 350 mm×10 mm) and closed cell foam board are set in the construction joints of lining, and two-component polysulfide sealant is inlaid on the surface.

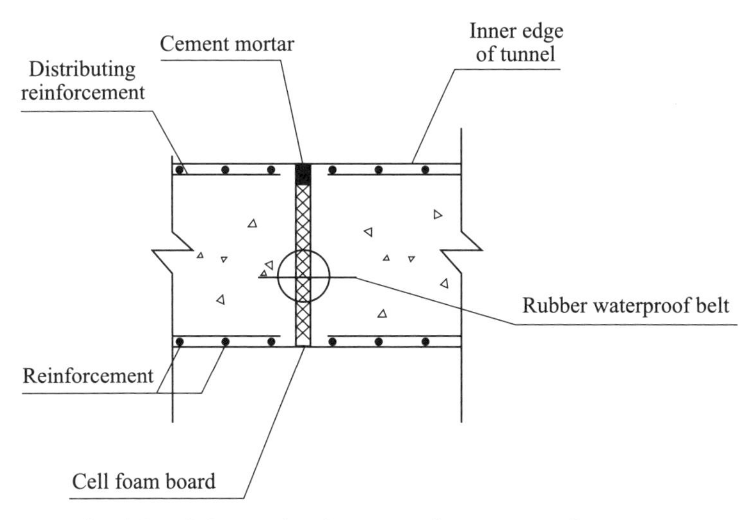

Fig. 1.7 Schematic diagram of waterproof structure

1.3 Structural characteristics

1.3.1 Force transfer process inside the MUAA lining

The mechanical characteristics of the MUAA lining are closely related to the load transfer mode of the annular anchor. Conventional anchors cause compressive stress in the concrete by making use of the compression from the anchoring end and the tensioning end, while the unbonded prestressed annular anchor uses the free anchor head which combines the anchoring end and the tensioning end to seal up the curved anchors, by means of "hoop effect". The annular tension of anchors is transformed into the radial load acting on the interface between anchors and lining, so as to produce precompression stress of concrete (Fig. 1.8).

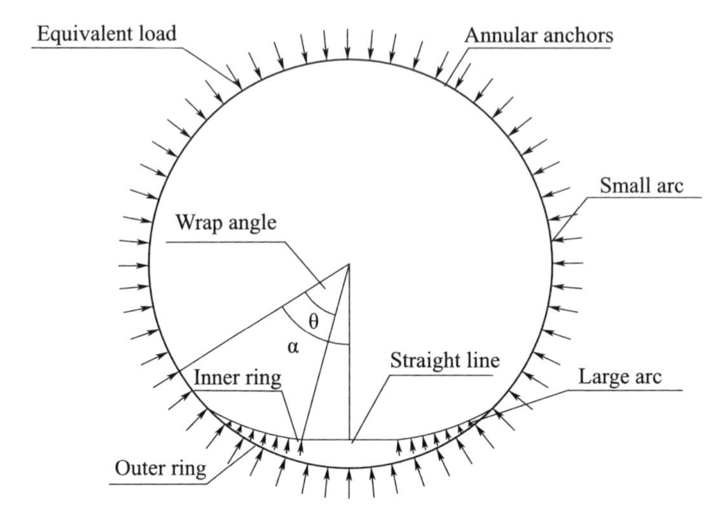

Fig. 1.8 Schematic diagram of unbonded annular anchors load in MUAA lining

In order to realize the load transfer within the lining by the unbonded annular anchor, the prestressing system of the structure is applied in the following approaches:

(1) The steel strand is wrapped with polyethylene case, and the internal gap is filled with grease or oil wax to reduce the friction of the anchors. The anchors are bundled, bent, and then bound to the inside of the conventional steel bar or set up against the steel bar.

(2) Anchorage block-out is reserved, and the anchoring end and the tensioning end of the anchors is fixed on the same anchor plate in it.

(3) Pour the lining concrete and cure to the design strength, use center hole jack on the annular anchor for grade tensioning. After there are applying tensile force in the annular anchor, tightly case the concrete near the hoop, so that the concrete compressive stress is created.

(4) Anchor the anchoring end and tensioning end to the free anchor head, and use the casing and epoxy resin to protect the exposed anchors against corrosion.

(5) Backfill micro-expansion concrete in the anchorage block-out, so

that the anchorage block-out has got a certain prestress, and the annular anchor and concrete can form a smooth and closed stressed structure.

When the MUAA lining is prestressed, it can bear the load of internal water, so as to realize anti-crack and anti-seepage. Moreover, this kind of active prestressed structure, unlike grouted prestressed lining, does not need the surrounding rock to provide reaction force, and can greatly reduce the proportion and action of surrounding rock in load sharing.

1.3.2 Combined bearing effect of surrounding rock and the MUAA lining

For the MUAA lining in the pressure tunnel, under the combined action of prestress, internal water pressure, external water pressure, dead weight stress and surrounding rock pressure, the combined bearing characteristics of surrounding rock and lining are more complex, and the interaction between surrounding rock and lining will change in different stages of construction period, operation period and maintenance period.

During the construction period, the surrounding rock is unloaded. After the lining is applied, the lining and the surrounding rock are squeezed together to bear the pressure of the surrounding rock. As the prestress is applied, annular anchor is rapidly tensioned by the jack, the lining inside shrink, and the upper part of the lining and surrounding rock release gradually, with the increase of tension, the gap opening between the surrounding rock and the lining gradually increases, and then, only the lower half of the lining is in contact with the surrounding rock, and the lining only bears self-weight stress, the annular anchor prestress and the supporting force of the lower surrounding rock, and no longer bears the load; with the backfilling and grouting behind the lining, the surrounding rock wraps the lining, and the two closely fit together to bear the load. Then, the contact pressure will gradually increase with the release of the surrounding rock stress.

During the operation period, the MUAA lining does not crack under internal water load in principle. Unlike limit design lining crack, the MUAA lining will not be separated from the surrounding rock due to the excessive seepage force of the internal and external water heads. Under the combined action of various pressures, the surrounding rock and the lining will squeeze and jointly bear the load. The load sharing between the two is related to the strength of the surrounding rock and the lining. When the prestress of the lining is sufficient, the load shared by the surrounding rock is slight.

During the maintenance period, with the dissipation of the internal water pressure, the MUAA lining shrinks again. The surrounding rock previously in the plastic state has residual deformation, and there will be a small gap between the lining and the surrounding rock after unloading, and the two are partially detached. The lining is mainly affected by prestress, external water pressure and dead weight stress, while the surrounding rock mainly needs to bear the pressure of itself.

1.4 Key points of structural design

1.4.1 Winding mode of annular anchor

The winding mode of the unbonded prestressed annular anchor is closely related to the mechanical properties of the structure, which is mainly about the setting of the number of layers and loops of the annular anchor.

The number of layers is the number of rows of annular anchors along the lining thickness. The single-layer winding method is to arrange all annular anchor bundles on the outside of the lining without setting inner annular anchor bundles. There are many annular anchors in the same layer. Considering the difficulty of pouring and ramming concrete between annular anchor bundles, it is necessary to increase the space of annular anchors. Its advantages are full utilization of the annular anchor tensioning force, with larger prestress, and can be conducive to reducing the lining thickness and project cost. The double-layer winding method is to arrange the prestressed annular anchor bundles into two rows, the inner and the outer. The distance between the two rows of anchors is generally 10 cm to 15 cm. The two rows of annular anchor bundles are parallel to each other and tied to the inner sides of the conventional steel bars on both sides, which can reduce the amount of binding of the erection bars. Its shortcoming is that the inner annular anchor bundle is too close to the inner side of the lining concrete, so it contributes less to the prestress, and the concrete vibration puddling between the two layers of annular anchors is difficult, so the lining must be reserved with sufficient thickness. With three or more layers of annular anchors, the lining is too thick to be used.

The loop of annular anchor is the number of turns that the annular anchor circles along the annular direction of the lining. The force diagram of single-loop and double-loop is shown in Fig. 1.9. The single-loop winding annular anchor structure is simple and easy to construct. However, due to the prestress friction loss along the path, the prestress value of the tensioning end and the anchoring end is obviously larger, while the prestress value far from the tensioning end and the anchoring end is smaller. So the overall structure stress is asymmetrically distributed, which increases the bending moment on the lining to a large extent. The double-loop winding annular anchor structure is more complex, so it is necessary to ensure accurate positioning of the double-loop anchoring and interlapping, and the alternating sequence must not be wrong. Its advantage is uniform prestress distribution of the lining structure, and there is no lack of prestress at the back of the anchorage block-out. When the number of annular anchors loop reaches three, the prestress loss is large and the material utilization rate is too low.

Therefore, considering the structural design and material utilization, there are mainly four types of unbonded prestressed annular anchor winding methods suitable for engineering applications: single-layer single-loop, single-layer double-loop, double-layer single-loop and double-layer double-loop.

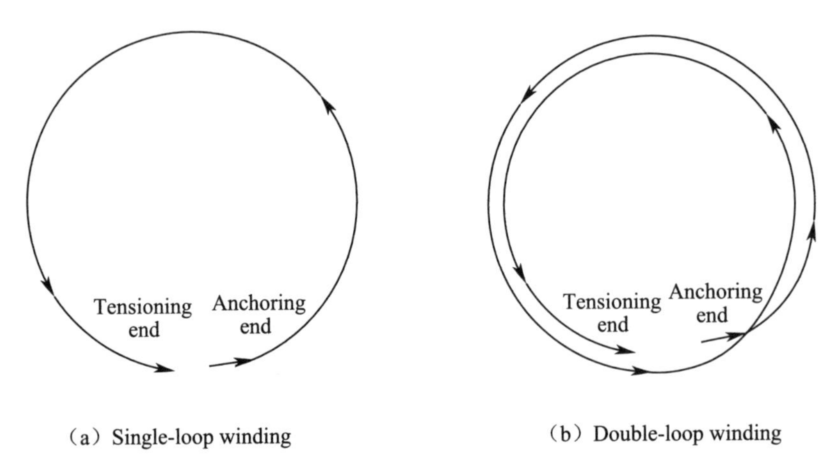

(a) Single-loop winding (b) Double-loop winding

Fig. 1.9 Winding mode of unbonded prestressed annular anchor

1.4.2 Position of anchorage block-out

Tensioning in-situ prestressed lining can only be carried out in the tunnel, so the anchorage block-out should be set up in advance for tensioning and anchoring the annular anchor before concrete lining is poured. The arrangement of anchorage block-out not only affects the distribution and range of the weak prestressed area of the lining, but also affects the backfilling density inside of the anchorage block-out. Therefore, it is one of the keys to determine the optimal position of anchorage block-out in the design of the MUAA lining.

When the anchorage block-outs are in the upper part of the tunnel, it is difficult to fix the tensioning equipment or to backfill the anchorage block-outs normally. Therefore, the anchorage block-outs are usually set at the bottom or side of the lining. When anchorage block-outs are in the bottom of the tunnel, the structure prestress distribution uniformity is fine, but it is easy to cause ponding, and constructors usually work at the bottom. The chipping of anchorage block-out before backfilling may leave a large amount of residue, so impurities and water retention are easily deposited in the anchorage block-outs, which is hard to be cleaned during backfilling.

Besides, when self-compacting concrete is used to fill the anchorage block-outs, the bottom of the latter is not conducive to discharge air bubbles from the concrete, so many bubbles are accumulated at the interface between the formwork and concrete, bringing a lot of holes on the surface of the backfilled concrete after the mold is removed. When anchorage block-outs are cross arranged at the side of tunnel at ±45°, the prestress distribution of the upper half lining is uniform, while that in the lower half lining is not uniform. However, from the perspective of construction, it is a little inconvenient only to put up the anchorage block-out formwork, and other problems can be well solved. The cross arrangement at ±45°of the anchorage block-out side can facilitate construction.

In addition, the compactness of concrete backfilling in anchorage block-outs is of great significance on anchorage protection. Since anchorage block-outs are generally arranged in a tilted way, the formwork fixing mode cannot be realized, so only manual vibration pouring concrete can be adopted. However, the compactness of the backfilled concrete is difficult to be guaranteed because of the dense prestressing tendons and structural reinforcing bars in the narrow anchorage block-out. The bonding force between the backfilled concrete and the lining is generally very poor, and it is easy to crack, so the prestressed reinforcement is likely to be exposed. The anti-corrosion grease may leak out, endangering the lining safety and affecting the internal water quality. The backfilled concrete in the anchorage block-out is self-stressed, and the prestress can be generated in the backfilled concrete through the expansion of self-stressed concrete, which can improve the adverse stress of anchorage block-outs.

1.4.3 Lining thickness

Reducing the lining thickness of the pressure tunnel can not only reduce project cost, but also improve operation efficiency of the water conveying tunnel by increasing water cross section area. Because the prestressed annular anchor is bound in the inner lining of non-prestressed steel bars, when the lining thickness is very small, the quality of concrete pouring will be affected, so the lining must maintain at a thickness conducive to construction, the general lining thickness is 0.45 m to 0.80 m.

The annular-compressive stress in the lining decreases with the increase of lining thickness, that is, the thinner the lining, the more significant the prestressing effect. However, with the decrease of lining thickness, the annular-tensile stress generated by internal water pressure also increases correspondingly. However the thickness of the lining has no significant effect on the bearing capacity of the prestressed lining under the condition of overloading water pressure, mainly because the prestress of the lining used to resist the internal water pressure ultimately comes from the tension of the annular anchor, and the change of the lining thickness mainly affects the distribution of the prestress.

Besides, the winding mode of annular anchor also has a great influence on the selection of lining thickness. The single-layer double-loop prestressed steel strand wrapping mode of unbonded prestressed annular anchor can effectively reduce the lining thickness.

1.4.4 Annular anchor spacing

The MUAA lining converts the annular tension of the annular anchor into the pre-compressive stress of the lining concrete through tensioned steel strand to resist the radial stress of the tunnel water acting on the lining surface. After the prestressed annular anchor is tensioned, it extends outward along the center line of the annular anchor bundle, and the pre-compressive stress of the lining decreases gradually. Therefore, the spacing between annular anchors has a great influence on the prestressing effect of the lining. The spacing should not be set too large in order to prevent the uneven distribution of prestress in the longitudinal direction of the lining. In general, the value of prestress gradually increases with the decrease of the annular-anchor spacing, and the smaller the spacing, the more obvious the increasing trend. Although anchor spacing significantly changes prestress value, basically it does not affect the stress distribution pattern.

1.4.5 Shape of cross section

The inside of the MUAA lining is usually circular, and the outside of the lining is similar in shape to horseshoe or three-centered arch, which may be due to the overcut and undercut of surrounding rock. Therefore, the outside section of actual lining may be circle, horseshoe, or three-centered arch, as shown in Fig. 1.10. As the inside of the sections are nearly uniform, internal lining annular anchor arrangement can stay unchanged.

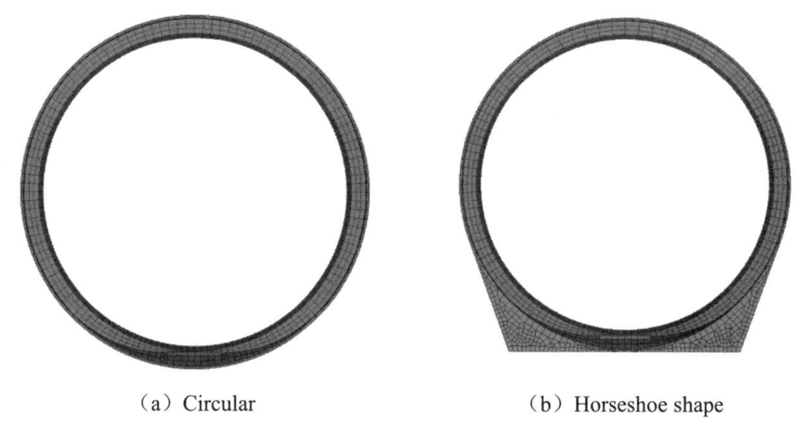

(a) Circular (b) Horseshoe shape

Fig. 1.10 Cross section shape of a typical MUAA lining

The surrounding rock pressure, internal and external water pressure and the tension of the annular anchor lining are all almost centrosymmetrically distributed. The cross section of the circular lining is typically centrosymmetric. Therefore, the overall stress distribution uniformity of the circular lining is better than that of other cross sections. However, with internal water pressure, resistance in the internal water ultimately comes from the prestressed anchors, with the same cable arrangement type, the ability to resist internal water pressure is largely independent of the type of hole. However, where the lining is weak near the anchorage block-outs, the structure may be safer for the horseshoe or three-centered lining is slightly thicker at the bottom. Considering that horseshoe or three-centered arch lining requires a significant increase in the amount of concrete, circular lining is more cost-effective.

References

CAO R L, WANG Y J, ZHAO Y F, et al, 2016. Study on numerical modeling method of the prestressed tunnel lining with unbonded curve anchored tendons[J]. *Journal of China Institute of Water Resources and Hydropower Research,* 14(6): 471-477.

CAO R L, WANG Y J, WANG X G, et al, 2019. Mechanical properties of pre-stressed linings with unbonded annular anchors under high internal water pressure based on large-scale in-situ tests[J]. *Chinese Journal of Geotechnical Engineering*, 41(8): 1522-1529.

WANG Y J, CAO R L, PI J, et al, 2020. Mechanical properties and analytic solutions of prestressed linings with unbonded annular anchors under internal water loading[J]. *Tunnelling and underground space technology*, 97(3): 1-10.

CAO R L, ZHAO Y F, WANG Y J, et al, 2021. Structural and Mechanical Characteristics of Double-Hoop Unbonded Annular Anchor Lining[J]. *ACI Structural Journal*, 118(3): 61-70.

KIRCHHERR, JULIAN, 2017. Conceptualizing Chinese engagement in south-east Asian dam projects: evidence from Myanmar's Salween River[J]. *International Journal of Water Resources Development*, 34(5): 812-828.

MATT P, THURNHERR F, UHERKOVICH I, 1978. Prestressed concrete pressure tunnels[J]. *Water Power and Dam Construction*, 1987(5): 23-31.

NAGAMOTO, TAKAYUKI, YONEDA, 2008. Large P & PC segment works: construction works for rainwater storage under Osaka International Airport[J]. *Underground and Tunnel*, 39(8): 581-589.

PI J, WANG Y J, CAO R L, et al, 2018. Innovative loading system for applying internal pressure to a test model of prestressed concrete lining in pressure tunnels[J]. *Journal of Engineering Research*, 6(2): 24-44.

SHEN F S, LIU X, 2003. Long and deep diversion tunnels of 1st stage project in west route of South-North Water Transfer Project[J]. *Chinese Journal of Rock Mechanics and Engineering*, 22(9): 1527-1532.

SIMANJUNTAK T D Y F, MARENCE M, MYNETT A E, et al, 2014. Pressure tunnels in non-uniform in situ stress conditions[J]. *Tunnelling and Underground Space Technology*, 42(5): 227-236.

TATE E L, FARQUHARSON F A K, 2000. Simulating reservoir management under the threat of sedimentation: the case of Tarbela dam on the river indus[J]. *Water Resources Management*, 14(3): 191-208.

TRUCCO G, ZELTNER O, 1978. Grimsel-Oberrar pumped storage system[J]. *Water Power & Dam Construction*, 1978(2): 351-357.

Chapter 2
Mechanical characteristics of the MUAA lining during the process of tension

2.1 Introduction

To reduce the pressure of the surrounding rock and make the lining resist high internal water pressure by relying only on its own prestressing, a high-strength anchor was applied in the pressure tunnel to form an active prestressed anchor. In the US, the underground water pipeline network adopted a prestressed concrete cylinder pipe (PCCP) (Hajali et al, 2016), but the prefabricated PCCP can only be used in an open tunnel excavation on a road instead of in a subsurface excavation. In Japan, the technology of assembling prestressed segments in the construction of underground drainage projects was put forward, but the prestressed segments are generally applied only in shield tunnels (Nishikawa, 2003). To achieve the field pouring of a prestressed lining, the technology of a prestressed anchor lining was proposed and applied in the surge tank in the Crimsel-Oberrar water pumped storage power station in Switzerland (Trucco & Zeltner, 1978). However, it is a structure of single annular anchor that can provide limited prestress with which is difficult to meet the needs of large cross section tunnels coping with high internal water pressure. Thus, in the construction of the desilting tunnel in the China Xiaolangdi multipurpose project and the Dahuofang water delivery tunnel, a new prestressed reinforced concrete lining with unbonded annular anchors has been developed (Kang et al, 2006; Kang et al, 2014). Furthermore, the new type of prestressed lining with a double-hoop unbonded annular anchor was proposed through construction of the long-distance water delivery tunnel for the water transfer project from the Songhua River (built in 2016-2020). Engineering practices show that prestressed lining with annular anchors has advantages of low prestress loss and uniformly distributed inward pressure, while, it also shows complex stress and strain mechanism between the annular anchor and the lining concrete. Thus, it is relatively difficult in construction due to its complicated structure (Cao et al, 2016). In view of the existing uncertainty, a somewhat large value of prestress load is generally utilized for improving safety, which results in local lining cracking nearby block-outs during operation of Xiaolangdi Desilting Tunnel (Lv et al, 2009).

The past long-term studies have gradually revealed the mechanical properties of the traditional prestressed structure and established the structural calculation theory (Mander et al, 1988; Ahn et al, 2010; Wan et al, 2002). In recent years, with the progress in engineering mechanics, numerical simulation and monitoring technology, research on the mechanical properties of prestressed structures has made new breakthroughs. For example, Park established a structural strength calculation method based on strain shear strength (Park et al, 2013). Lou proposed a theory of nonlinear time-varying structural analysis (Lou et al, 2013). Sousa expounded the prestressed structural loading characteristics under construction and long-term mechanical properties in the operation stage with the statistics and analysis of monitoring data (Sousa et al, 2013). However, the conventional prestressed structure squeezes between the pretension and anchor end of the beam to directly generate the prestress (Zhang et al, 2018). The circumferential tension of the anchor is converted to the radial load on the interface between the anchor and the lining via the hoop effect, by which compressive prestress is generated in concrete. These two kinds of structures are obviously different from each other, so the research results of traditional prestressed anchor structures are not suitable for the MUAA lining. Some researchers (Nagamoto et al, 2008; Lee et al, 2015) have shifted their focus on the annular anchor structure and explored the deformation rule of prestressed steel tubes, concrete pipe structures and methods for determining the stiffness and bearing capacity of prestressed anchor segments, but the MUAA lining adopted the double-hoop methods, including embedded anchorage block-out and stepwise tensioning (Cao et al, 2016). The annular anchors of MUAA lining show quite different force transfer mode compared with conventional prestressed anchor structure. The mechanical concept of the conventional prestressed annular anchors at anchoring end and tensioning end is clear, that is, the squeezing action between two ends generates the compressive stress in concrete (Fahimifar and Soroush, 2005; Showkati et al, 2016). The circumferential tension of anchor is converted to the radial load on the interface between anchor and lining via the "hoop effect", in which way compressive prestress is generated in concrete. Nowadays, the related researches on prestressed lining with annular anchors do not link of the load transfer mode with the mechanical properties closely. Therefore, the previous research results are not applicable to MUAA lining, and it is not conducive to revealing its mechanical properties and establishing the calculation method of mechanical state.

In view of this, a large in-situ tension test of the MUAA lining was carried out for the first time based on the long-distance pressure tunnel under construction. Under the premise that the mechanical boundary condition and loading mode of the lining structure in the test are consistent with the actual engineering, the distribution rules of annular anchor tension and lining stress have been obtained. At the same time, the superposition effect of the annular anchor tension, spatial distribution pattern of the lining prestress and potential destroyed mode of the structure were deeply explored with theoretical analysis methods.

2.2 In-situ test of the MUAA lining

2.2.1 Structure

The structural design scheme of the MUAA lining in-situ test is shown in Fig. 2.1. The inner diameter of the lining is 6.9 m, the outer straight diameter is 7.8 m, the wall thickness is 0.45 m, and the length is 3.0 m. Four prestressed anchors are laid in parallel to form a bundle and then bent according to the preset strength. The annular anchor starts winding from the anchor end to the outside of the lining, wraps around 720° along the lining and is fixed at the tension end. The ends of the anchor and tensioning are placed in the same radial load head for closing the anchor to form an annulus, and the anchorage block-out is interlocked at ±45° in the inner side of the lining of the lower half annulus. It is not only used for burying the radial load head and the anchor system but also provides working space for later tensioning construction [Fig. 2.1 (c)]. To ensure the uniformity of the structure prestress, the integral lining and annular anchor are distributed symmetrically in the center except in the vicinity of the anchorage block-out. This study, however, solely includes one lining section measuring 7.8 m in diameter, more sections and different lining conditions should be introduced in the future study.

The parameters of pressure tunnel, surrounding rock and MUAA lining in the in-situ testing are listed in Table 2.1. The construction process of the MUAA lining includes anchor installation [Fig. 2.2(a)], anchorage block-out manufacturing [Fig. 2.2(b)], lining casting [Fig. 2.2(c)] and buried installation of the anchor device [Fig. 2.2(d)].

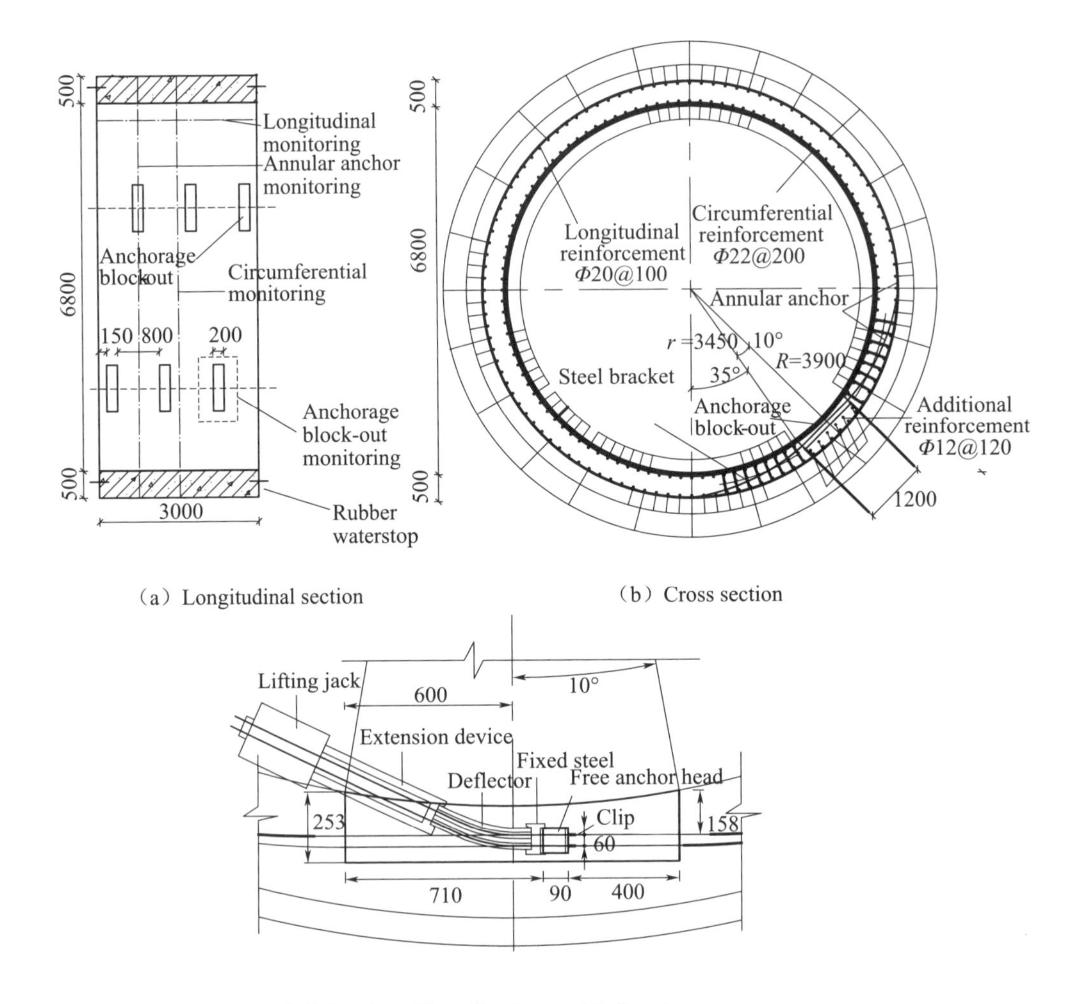

(a) Longitudinal section

(b) Cross section

(c) Local position of anchorage block-out

Fig. 2.1 The Structural design drawings of MUAA lining in the in-situ testing (unit:mm)

(a) Annular anchor installation

(b) Anchorage block-out manufacture

Fig. 2.2(1) Construction process of MUAA lining in the in-situ testing

(c) Lining pouring

(d) Anchorage device Installation

Fig. 2.2(2) Construction process of MUAA lining in the in-situ testing

Table 2.1 Parameters of pressure tunnel, surrounding rock and MUAA lining in the in-situ testing

Item	Surrounding rock					MUAA lining			Tunnel	
	Elasticity modulus	Poisson's ratio	Angle of internal friction	Cohesion	Initial stress	Elasticity modulus	Poisson's ratio	Thickness	Inside radius	Internal pressure
Symbol/ Unit	E_r/GPa	μ_r	φ/(°)	c/MPa	q_0/ MPa	E_c/GPa	μ_c	d/m	r/m	P_0/MPa
Value	0.3	0.38	23	0.05	0.23	35.01	0.20	0.45	3.45	0.60

2.2.2 Test materials

Lining concrete uses ordinary silicate cement, whose strength grade is C40, elastic modulus 32.5 GPa, slump 179 mm, and water-binder ratio 0.32. See Table 2.2 for concrete material consumption per cubic meter.

Table 2.2 Quantity of concrete material

Material	Cement	Fly ash	Sand	Stone	Small stone	Water	Water reducer
Quantity / (kg/m^3)	352	88	695	576	384	141	141

The physical and mechanical parameters of the unbonded prestressed anchors are shown in Table 2.3. Four anchors are placed side-by-side to form a bundle of annular anchors, and the spacing of each annular anchor bundle is 500 mm. Its plane is perpendicular to the direction of the pressure tunnel. The reinforcement is laid in advance and the feature points tagged to control the position of the annular anchor installation and ensure that the annular anchor curve is smooth without dislocation and intersection.

Table 2.3 Parameters of annular anchor in the in-situ testing

Parameters	D/mm	f_{ptk}/MPa	A_p/mm^2	f_{ptk}/kN	E_s/GPa	f_p/kN
Value	15.24	1860	140	260.5	195	195.4

2.2.3 Test monitoring

The annular anchor tension, lining prestress and structural stress near the anchorage block-outs were mainly monitored for further analysis of the mechanical characteristics of the MUAA lining. The monitoring positions are shown in Fig. 2.1(a), and the sensor layout scheme is shown in Fig. 2.3.

A magnetic flux sensor [Fig. 2.3(a)] was used to monitor the annular anchor tension with an accuracy of 0.01 kN. The sensors at the pretension and anchor ends are numbered C1 and C7, respectively, and the anchor round radius change position and the midpoints of sensors C2-C6 are shown in Fig. 2.3(b). A fiber Bragg grating sensor [Fig. 2.3(c)] is embedded inside the lining to measure the prestress value of the concrete. The numbering of the sensors clockwise increased from FGS1 to FGS16, and the longitudinal numbering of these sensors in FGS17-FGS22 in sequence are shown in Fig. 2.3(b). The resistive strain gauges, which are numbered S1-S41, are arranged at characteristic points around the anchorage block-out [Fig. 2.3(d)] to monitor the local stress and strain [Fig. 2.3(e)], and the measuring accuracy is 0.02 με.

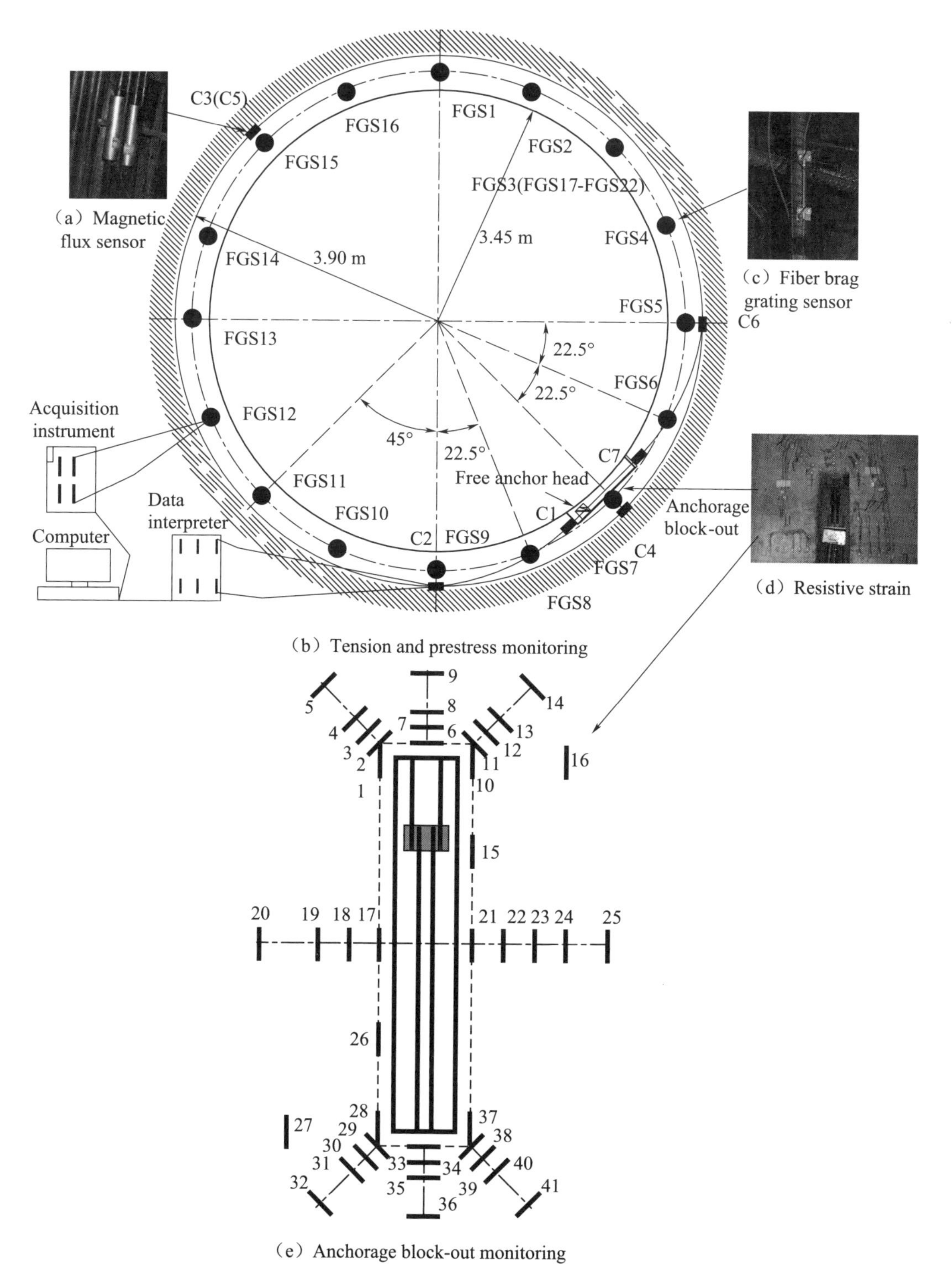

Fig. 2.3 Monitoring program plan and installation for MUAA lining

2.3 Tension of the annular anchor

The MUAA lining is tensioned by the lifting jack, and the elongation value of annular anchors under various loads is taken as the control standard of tension (Fig. 2.4). To allow sufficient time for the jack force to transfer from the tension end to the anchor end, controlling the loading time of each stage and the interval time between stages during in-situ testing is required. The tensioning of the annular anchor will sharply increase the prestress of the inner side of the concrete. If the prestress of the inner side of the concrete is greatly different from that of the adjacent concrete, it will produce serious shear stress on the inclined section. To reduce the shear stress in the process of annular anchor tensioning, the different steps of the tension method shown in Fig. 2.5 are adopted in the test. The principle is that the tension difference in any adjacent annular anchor shall not exceed 50% of the working load (f_p). In Fig. 2.5, 1#-6# are the numbered annular anchors, "Step1-2#-50%" denotes that the first step of the tension construction is to make the designed load of the tension of the 2# annular anchor reach 50%. Step7-2# 100% means that the 7th step of the tension construction is to make the design load of the tension of the 2# annular anchor reach 100%.

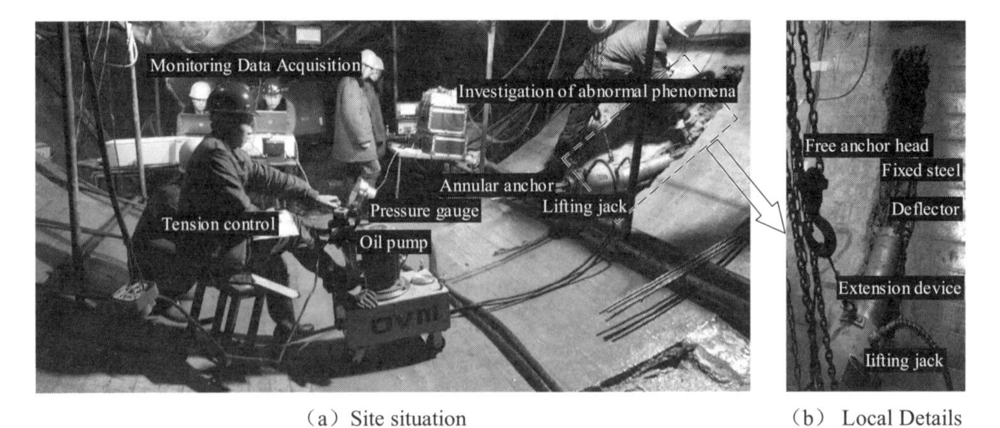

(a) Site situation (b) Local Details

Fig. 2.4 Information collection of in-situ test annular anchor lining tension and monitoring

The in-situ tension test of the MUAA lining and the monitoring information of the mechanical properties are shown in Fig. 2.6. Since the length of the annular anchor is up to 60 m, there is a noticeable time delay during the tension transfer inside the MUAA lining. Thus, in addition to the stepped loading, the anchoring should meet the following two requirements: the loading speed must be controlled within 20 kN/min and the tension must remain at least 10 min at the designed value.

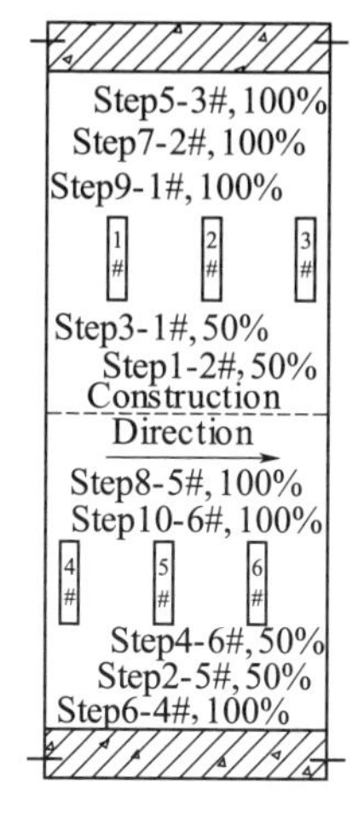

Fig. 2.5 Order of the prestressed annular anchor tension

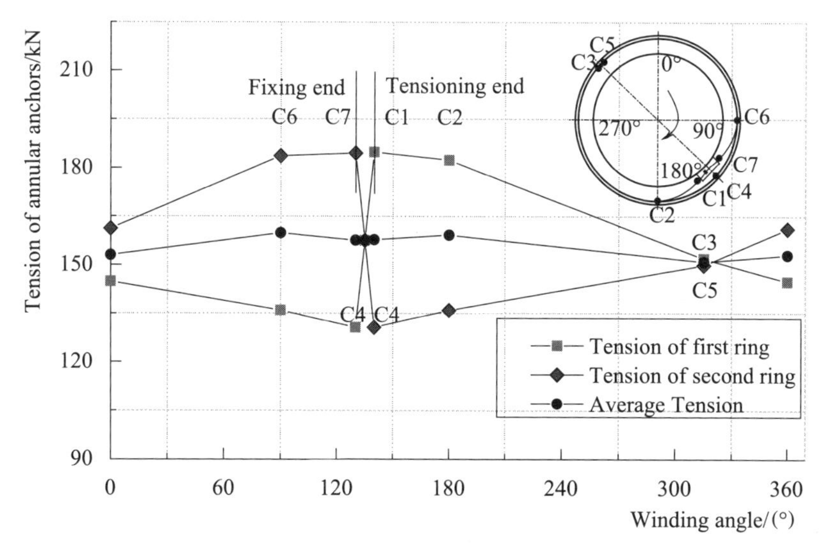

Fig. 2.6 Tension distribution along annular anchor

2.4 Superposition effect of the annular anchor tension

The lining prestress comes from the annular anchor, so the distribution of the annular anchor tension decides the quantity and uniformity of the prestress of the MUAA lining. The greater and closer the anchor tension of the potential annulus is, the more is favorable the stress of the lining structure.

Fig. 2.6 shows the distribution curve of the annular anchor tension around the hole. The tensions at the measuring points from C1 to C4 or from C7 to C4 basically show symmetric attenuation. And the smaller is the tension, the farther away are the points from the stressing end or fixed end of the annular. The minimum tension (130.61 kN) appears at the center of the annular anchors (C4). The design of one-layer double-hoop annular anchors make the location of maximum tension (C1, 184.90 kN and C7, 184.53 kN) close to the location of minimum tension (C4). The average tension at C1, C4 and C7 is 157.66 kN, which is only about 4.2% higher than the average tension (150.93 kN) at the farthest positions (C3, 151.97 kN and C5, 149.90 kN) of the anchorage block-out. Although the tension distribution along the annular anchors is still nonlinear, the difference is acceptable. Thus, conclusions can be drawn that the use of double-hoop winding mode can ensure approximately uniform distribution of prestress.

Since the annular anchor is closed with the radial load head, the tension end is very close in the anchor end (C1, 184.90 kN) and the tension end (C7, 184.53 kN). Compared with the conventional anchor, the force in the annular anchor is no longer large at the tension end and small at the anchor end, and the tension difference between the two ends is now small. The distribution characteristics of prestress loss of conventional anchor and annular anchor of the same length are shown in Fig. 2.7. For the same length of anchor, the maximum prestress loss (L_{max}) of single-hoop annular anchor is half of that of conventional anchor, and the maximum prestress loss (L_{max}) of double-hoop annular anchor is one quarter of that of conventional anchor. The total prestress loss of single-hoop annular anchor and double-hoop annular anchor is almost equal, but the distribution of prestress loss of double-hoop annular anchor is very uniform.

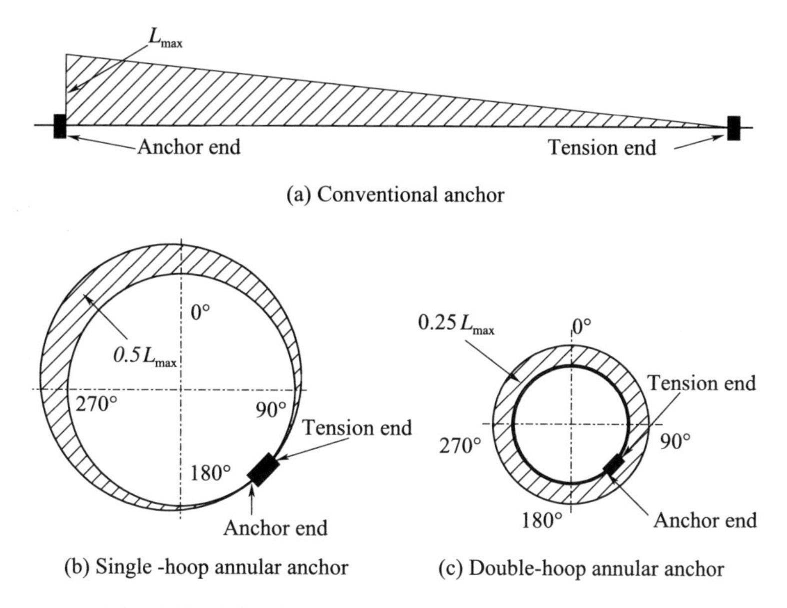

Fig. 2.7 Distribution characteristics of prestress loss along the anchor

In addition, the designed value of the annular anchor tension is 195.40 kN, the tensile end tension is 184.90 kN, and the tensile anchorage loss is 9.46%. In reality, the actual average tension of the annular anchor is 156.10 kN, and the prestress loss along the way is 9.66%.

2.5 Spatial distribution pattern of the MUAA lining

2.5.1 Annular prestress

The whole process of the on-site annular anchor tensioning test consists of ten steps (Fig. 2.5). The growth curve of the circumferential prestress on the corresponding characteristic points of the lining center section for each step is shown in Fig. 2.8.

The final average prestress of the right and left part of the MUAA lining is 6.25 and 5.91 MPa, respectively, with a difference of only 0.34 MPa. Thus, the MUAA lining prestress of the two prestressed annuluses are distributed symmetrically when both the laying structure and the arrangement of the annular anchor satisfy the geometric symmetry condition, especially when the superposition values of the anchor tension are relatively close to each other (Fig. 2.6).

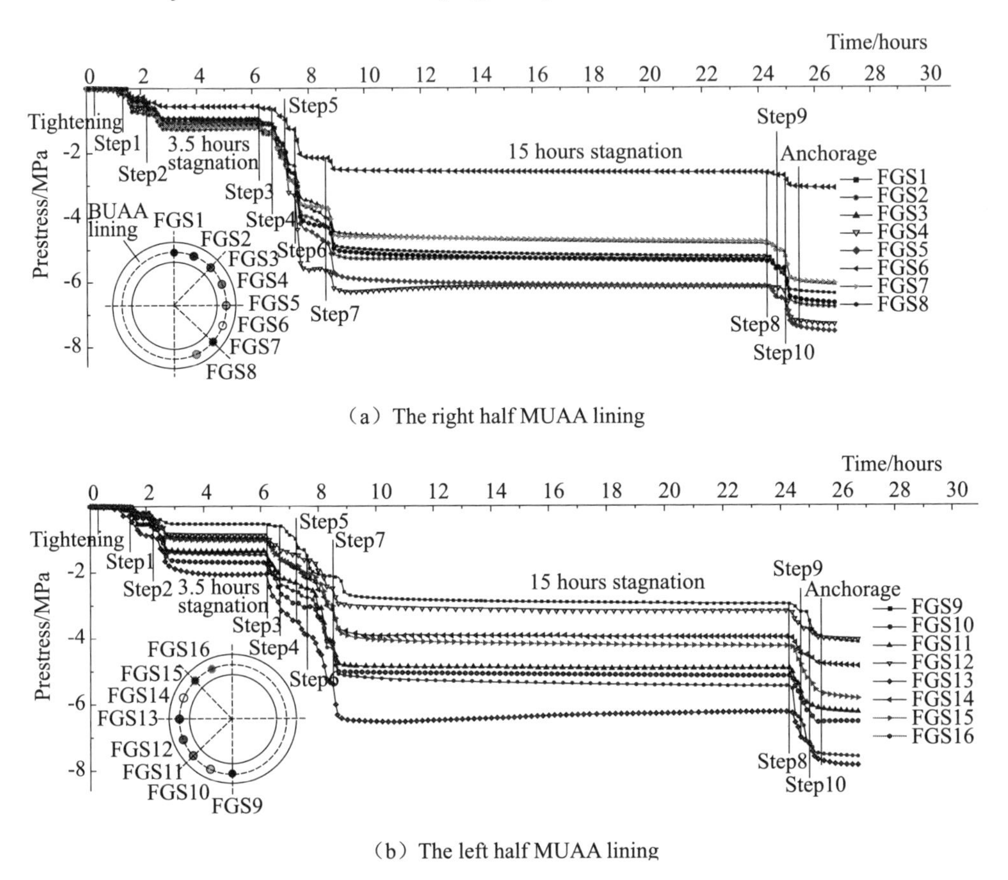

(a) The right half MUAA lining

(b) The left half MUAA lining

Fig. 2.8 The growth rule of the prestress during the process of annular anchor tension

When the tension load of the 5# circular anchor, which is nearest to the lining center section, increases from zero to 50% (Step 2) and from 50% to 100%, the increased prestress value accounts for 10.84% and 8.70%, respectively, for a total of 19.54%. When the tension load of the 3# circular anchor, which has the longest distance from the lining center section, reaches 100% (Step 5), the prestress value increases by 10.51%, which is only half of that value of the 5# circular anchor. It is obvious that the prestress value from the circular anchor is closely related to its location. The statistical analysis results from the prestress contribution of different locations in the monitored section are shown in Fig. 2.9. The quadratic function relationship between them is as follows (r^2=0.9541):

$$\Delta\sigma = 1.372x^2 + 9.573x + 3.153 \quad (2.1)$$

where: $\Delta\sigma$ refers to the prestress increment (MPa); x refers to the distance from the circular anchor to the lining center section (m).

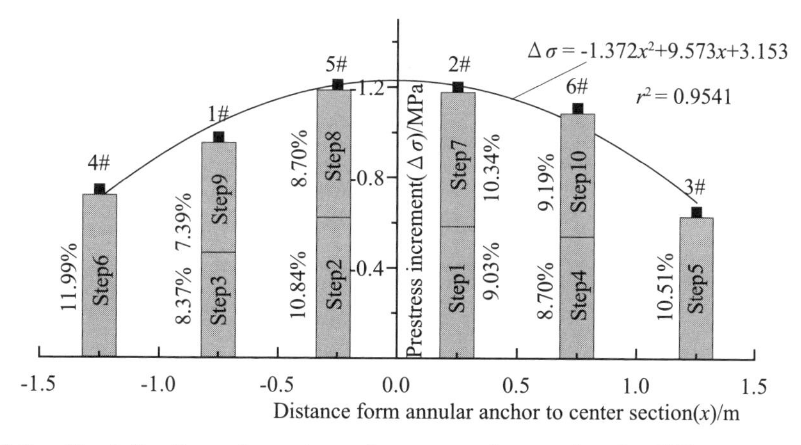

Fig. 2.9 Contribution of prestress from annular anchor in different positions

Fig. 2.10 shows the distribution curve of the prestress of the MUAA lining with an average value of 6.09 MPa. The maximum prestress value is 7.87 MPa, which appears at the left surround angle of 270° of the lining, and the prestress at the right symmetrical position base (surround angle of 90°) is relatively large, which is 7.55 MPa. The main reason for the large stress in these two parts is the influence of the anchorage block-out. Anchorage block-outs are arranged in the range of angles 122.5°- 135° and 225°- 247.5°, which will cause sudden changes in the force in the vicinity, during which the bending moment is produced, except from the original stress. The superposition of the original stress and bending moment results in greater local stress. At the same time, the above sudden changes will also cause the stress on the back of the anchorage block-outs (surround angle from 112.5° to 247.5°) to be less, with a minimum value of 3.12 MPa, which is only 51.2% of the average stress. Thus, the surrounding area of the anchorage block-out is regarded as the weak area in the structure with sudden changes in force. The following sections analyze the mechanical properties of the anchorage block-out.

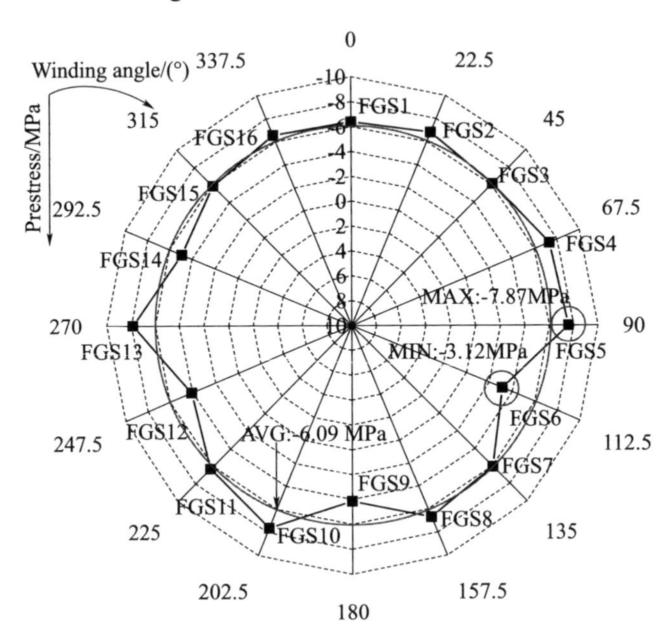

Fig. 2.10 Average tension of annular anchors during loading process

2.5.2 Longitudinal prestress

To further determine the relationship between the annular anchor position and prestress to analyze the uniformity of the longitudinal prestress, the growth curve of longitudinal prestress in Fig. 2.10 is charted according to the monitoring results.

The longitudinal prestress distribution curve is closely related to the MUAA lining boundary. The stress peak of the annular anchor near the boundary is larger when the prestress decrease along the longitudinal direction is more obvious. The prestress distribution curves of the annular anchors (1# and 6#, 2# and 5#, 3# and 4#) in symmetrical positions are very close to each other and can be described by the cubic function [$\sigma_1(x)-\sigma_6(x)$], which can be expressed by the matrix in Formula (2.2):

$$\begin{pmatrix} \sigma_1(x) \\ \sigma_2(x) \\ \sigma_3(x) \\ \sigma_4(x) \\ \sigma_5(x) \\ \sigma_6(x) \end{pmatrix} = \begin{pmatrix} -0.349 & 0.037 & 1.152 & -1.038 \\ 0 & 0.420 & -0.259 & -1.361 \\ 0 & -0.393 & -1.058 & -0.732 \\ 0 & -0.391 & 1.032 & -0.716 \\ 0 & 0.329 & 0.299 & -1.259 \\ 0.423 & 0.068 & -1.242 & -1.080 \end{pmatrix} \begin{pmatrix} x^3 \\ x^2 \\ x \\ 1 \end{pmatrix} \tag{2.2}$$

The geometric area formed by the projection of each prestressed distribution curve to the horizontal axis can be expressed as:

$$A_i(x) = \int_{x_1}^{x_2} \sigma_i(x)\mathrm{d}x \tag{2.3}$$

where: x refers to the distance from the annular anchor to the lining center section (m); x_1 and x_2 refer to the left and right boundaries, respectively, of the longitudinal lining, whose values in this test are taken as –1.5 m and 1.5 m.

Although the curves are different, the integral area of the curves $[\sigma_i(x)]$ is basically the same and close to 2.3 MPa·m, which indicates that the sum of the prestress increases from the MUAA lining generated by each annular anchor is the same when the allocated weights are different in different positions.

The final prestress of the MUAA lining is the superposition of prestress generated by each annular anchor, and the stress superposition value is very close. The minimum value and maximum value of the prestressed concrete lining along the longitudinal annulus are 5.34 MPa and 5.68 MPa, respectively, and the relative difference between them and the average value (5.43 MPa) is within 5%. The distribution of the longitudinal lining stress is considered to be nearly uniform.

2.5.3 Stress concentration at the anchorage block-out

As mentioned above, the annular prestress distribution curve of the MUAA lining in Fig. 2.10 shows that the sudden change in lining stress and structural weakness are located near the anchorage block-out. For this reason, the changes in the concrete in this part are mainly monitored using a resistance strain gauge. Then, the monitoring results are drawn into a contour map with Surfer software (Fig. 2.11).

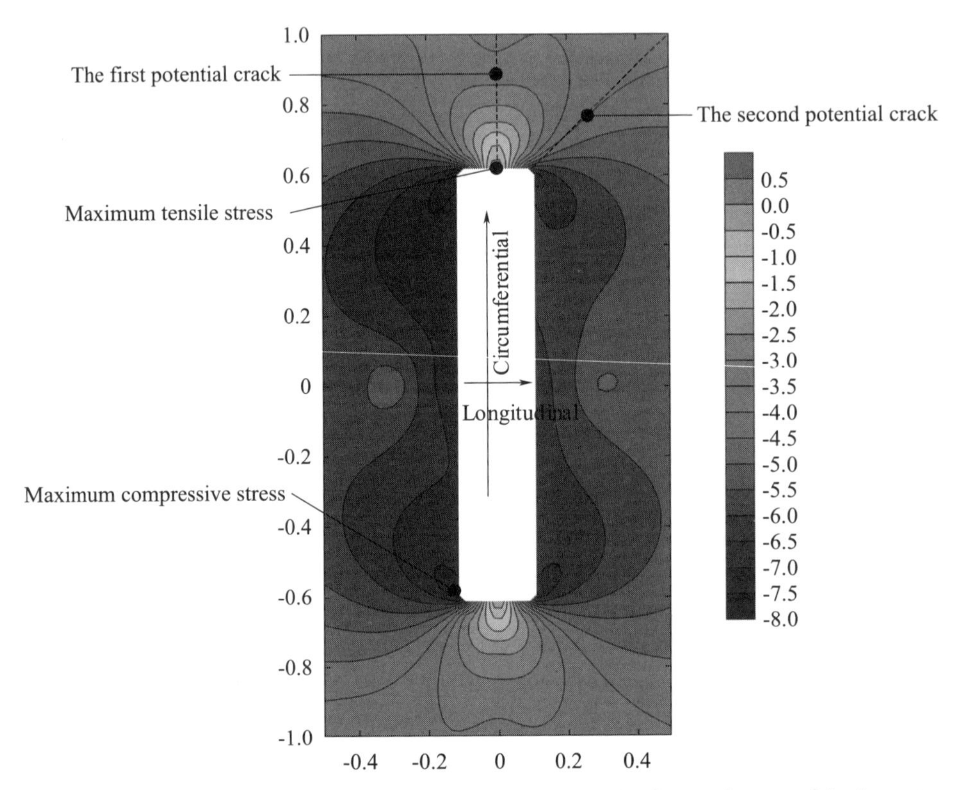

Fig. 2.11 Stress concentration and potential cracks in anchorage block-out.

The monitoring results reflect that the stress in the anchorage block-out and its vicinity area is obviously uneven. The compressive stress on the lining at both ends of the rectangular anchorage block-out is obviously large. The overall compressive stress is 6.50 MPa to 8.50 MPa, and the maximum compressive stress of this part is approximately 14 times that of the whole lining prestress (6.09 MPa). However, the prestress value of the lining at the end of the rectangular anchorage block-out annulus is small, and the tensile stress is 0.56 MPa. Although the tensile stress is less than the tensile strength of C40 concrete (1.71 MPa), it could still be the potential cracking location, forming the weak area of the prestress. If the annular anchor tension continues to increase, there will be two kinds of tension cracks around the anchorage block-out, as shown in Fig. 2.11. The first potential crack is perpendicular to both ends of the anchorage block-out and cracks deep in the concrete along the root of the circular anchor. The second potential crack might appear at the four corners of the rectangular anchorage block-out and extend inside the lining at 45°. Thus, it is necessary to strengthen the amount of reinforcement and compacted casting near the anchorage block-out, and the reinforcement should be enclosed into an annulus along the periphery of the anchorage block-out to prevent local damage of the MUAA lining.

2.6 Analytical method of structural force

In the process of tension, the MUAA lining shrinks to the inside of the tunnel and the surrounding rock and the upper half of the lining will be separated, but the lower surrounding rock only plays a role in supporting the lining, and the bearing capacity of the surrounding rock can be negligible. Fig. 2.12 shows that the stress on the MUAA lining is basically the same along the longitudinal direction, which conforms to the plane strain hypothesis of elastic mechanics. Then, the MUAA lining mainly bears internal water pressure load (P) and external prestress (P_p) under the action of the internal water pressure.

According to the mechanics principle of the tunnel, the mathematical model of this stress system can adopt the thick-walled cylinder under uniform internal and external pressure (Zareifard & Fahimifar, 2016), as shown in Fig. 2.13. Therefore, according to classic solutions (Lame's solution) of the stress distribution of thick-walled cylinders (Qayssar et al, 2014), the prestress of the lining (σ, MPa) can be obtained by formula derivation as follows:

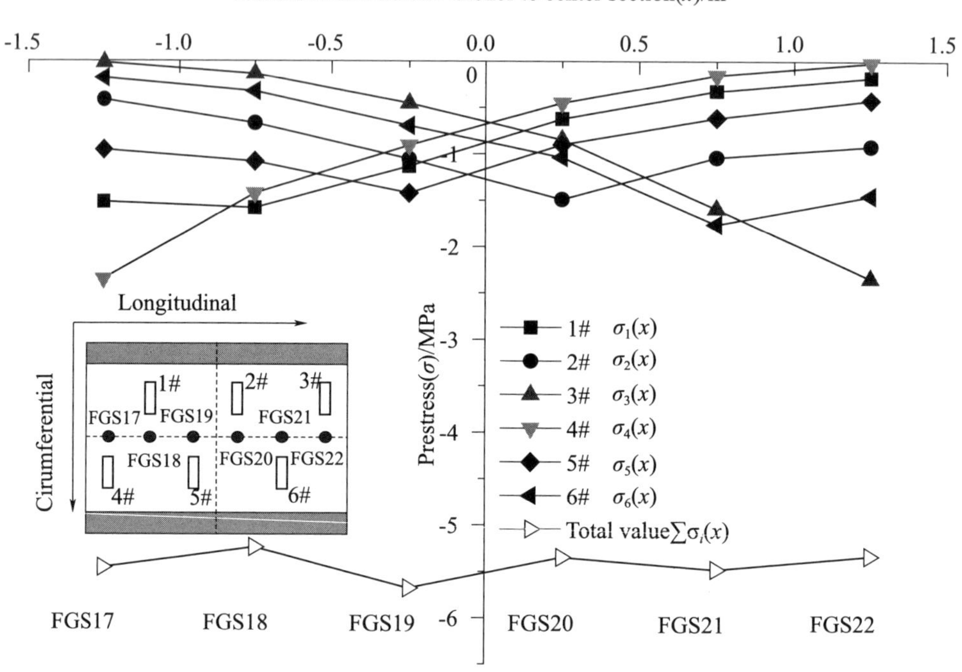

Fig. 2.12 Stress on the MUAA lining along the longitudinal direction

$$\sigma = \frac{P_0 a^2 - P_p b^2}{b^2 - a^2} + \frac{\left(P_0 - P_p\right) a^2 b^2}{\left(b^2 - a^2\right) r^2} \tag{2.4}$$

where: a refers to the inner radius of lining (m); b refers to the outer radius of lining (m); r refers to the distance from the stress position to the origin (m); P_p is the equivalent prestress generated by the annular anchor (MPa); and P_0 is the load of the internal water pressure (MPa).

If the MUAA lining has been strengthened at the weak part of the lining and the failure of that part is not taken into account, the following results (Table 2.4) can be obtained by analyzing the stress characteristics of the structure.

(1) Equivalent Prestress. In the test, the average prestress is 6.09 MPa with the tension of the MUAA lining. Therefore, the stress state of the structure is P_0=0 MPa, σ=6.09 MPa. Substituting into Formula(2.4), the equivalent prestress of 0.75 MPa can be obtained.

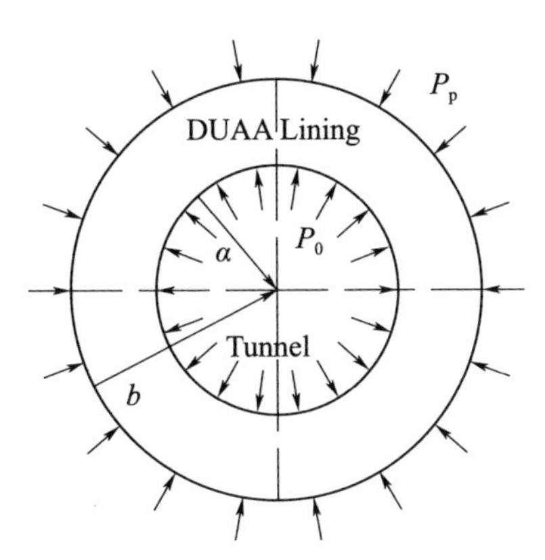

Fig. 2.13 Mathematical model of the MUAA lining

Table 2.4 Structural stress state of MUAA lining under different conditions

Stress state of MUAA lining	Internal water pressure P_0/MPa	Equivalent prestress P_p/MPa	Prestress of lining σ/MPa
Equivalent prestress	0	0.75	6.09
Optimal stress	0.79	0.75	0
Critical cracking load	1.02	0.75	1.71
Bearing capacity of structure	3.50	3.00	0

(2) Optimal Stress State. Under the loading condition of internal water pressure, if σ=0 MPa and P_p=0.75 MPa, P_0= 0.79 MPa can be obtained from Formula (2.4). That is, when bearing the internal water pressure load of 0.79 MPa, the prestress and internal water load can offset each other, the average stress of the lining is close to zero, and the stress state is optimal.

(3) Critical Cracking Load. The MUAA lining uses concrete with C40 strength grade in the test, and the design value of the tensile strength is 1.71 MPa. If σ=1.71 MPa and P_p=0.75 MPa, P_0=1.02 MPa can be obtained from Formula (2.4). Therefore, in the test, the bearing load of the internal water pressure under the critical cracking state of the MUAA lining is 1.02 MPa.

(4) Bearing Capacity of the Structure. The lining prestress comes from the annular anchor. If the interval distance between the annular anchors is halved and the number of annuluses is doubled, the equivalent prestress (P_p) theoretically can be increased by four times to 3.0 MPa. Generated from Formula (2.2), the bearing load of internal water carried by the lining will exceed 3.5 MPa, i.e. 350 m hydraulic pressure. Practically, if the design value of the internal water bearing load changes, such parameters as annular anchor tension, quantity and interval distance can be flexibly adjusted to provide the corresponding prestress. Therefore, even when considering the load shearing effect of the surrounding rocks, the MUAA lining can meet the supporting requirements of most pressure tunnels.

2.7 Discussion and conclusions

The concrete tunnel linings actively prestressed by posttensioning unbonded annular anchors still have a considerable dilemma to deal with, such as the distribution of annular anchor tension as well as the production mechanism of inward pressure on the lining induced by the anchor. The structural and mechanical characteristics of a new type of prestressed reinforced concrete lining with double-hoop unbonded annular anchors was evaluated and compared to that of conventional prestressed structure strengthened with unbonded anchors; there is a significant lack of such data in the current literature. The results obtained provide more details on the annular anchor tension, the spatial distribution patterns of the prestress and the potential failure mode of the structure by closely considering the mechanical properties and the prestress increasing together. Additionally, new analytic solutions to the lining stress by annular anchors and internal water pressure are proposed, which are verified by the in-site test results.

With an in-situ test and theoretical analysis, the structural characteristics and mechanical properties of the new type of MUAA lining have been identified. Without considering the bearing conditions of the surrounding rocks, the MUAA lining provides a uniform circular and longitudinal prestress through the active role of constraints, which can effectively resist the high internal water pressure force and satisfy the supporting requirements of most pressure tunnels. The loss of prestress can be greatly reduced by enclosing the anchor by using the radial load head, and the very close superposition value of tension from the double annular anchor can promote the uniformity of prestress of the MUAA lining.

Although the effect of the annular anchor on the annulus stress of the lining varies with the location, the increase in the prestress is the same. The growth curve of longitudinal prestress can be described by a cubic function when the annular anchor is tensioned. The weak link of the MUAA lining is the anchorage block-out, whose two ends and four corners are prone to tensile cracks, so the reinforcement must be strengthened to reduce the possibility of local cracking.

References

AHN J H, JUNG C Y, KIM S H, 2010. Evaluation on structural behaviors of prestressed composite beams using external prestressing member[J]. *Structural Engineering and Mechanics*, 34(2): 247-275.

CAO R L, WANG Y J, ZHAO Y F, 2016. Study on numerical modeling method of the prestressed tunnel lining with unbonded curve anchored tendons[J]. *Journal of China Institute of Water Resources and Hydropower Research*, 14(6): 471-477.

FAHIMIFAR A, SOROUSH H,2005. A theoretical approach for analysis of the interaction between grouted rock bolts and rock masses[J]. *Tunneling and Underground Space Technology*, 20: 333-343.

HAJALI M, ALAVINASAB A, & SHDID C A, 2016. Structural performance of buried prestressed concrete cylinder pipes with harnessed joints interaction using numerical modeling[J]. *Tunnelling & Underground Space Technology*, 51(1):

11-19.

KANG J F, LIANG Y H, ZHANG Q C, 2006. 3-D finite element analysis of post-prestressed concrete lining[J]. *Journal of Tianjin University Science and Technology*, 39(8): 968-972.

KANG J F, SUI C E, WANG X Z, 2014. Structure optimization of the prestressed tunnel lining with unbonded circular anchored tendons[J]. *Journal of Hydraulic Engineering*, 45(1): 103-108.

LEE Y, LEE E T, 2015. Analysis of prestressed concrete cylinder pipes with fiber reinforced polymer[J]. *KSCE Journal of Civil Engineering*, 19: 682-688.

LOU T, LOPES S M R, LOPES A V, 2013. Nonlinear and time-dependent analysis of continuous unbonded prestressed concrete beams[J]. *Computers & Structures*, 119(4): 166-176.

LV L, YANG Y, TANG W, 2009. Treatment for oil and water penetration in anchorage channel of san discharge tunnel Xiaolangdi key water control project[J]. *Chinese Building Waterproofing*, 2009(2): 36-38.

MANDER J A B, PRIESTLEY M J N, 1988. Theoretical stress-strain model for confined concrete[J]. *Journal of Structural Engineering*, 114(8): 1804-1026.

NAGAMOTO, TAKAYUKI, YONEDA, 2008. Large P & PC segment works: construction works for rainwater storage under Osaka International Airport[J]. *Tunnels and Underground*, 39(8): 581-589.

NISHIKAWA K., 2003. Development of a prestressed and precast concrete segmental lining[J]. *Tunnelling & Underground Space Technology*, 18(2): 243-251.

PARK H G, KANG S, CHOI K K, 2013. Analytical model for shear strength of ordinary and prestressed concrete beams[J]. *Engineering Structures,* 46(1): 94-103.

QAYSSAR S M, MOHAMMAD T, PRABHAT K S, 2014. Analysis of a thin and thick-walled pressure vessel for different materials[J]. *International Journal of Mechanical Engineering and Technology*, 10(5): 9-19.

SHOWKATI A, MAAREFVAND P, HASSANI H, 2016. An analytical solution for stresses induced by a post-tensioned anchor in rocks containing two perpendicular joint sets[J]. *Acta Geotechnica*, 11(2): 415-432.

SOUSA H, BENTO J, FIGUEIRAS J, 2013. Construction assessment and long-term prediction of prestressed concrete bridges based on monitoring data[J]. *Engineering Structures*, 52(9): 26-37.

TRUCCO G, ZELTNER O, 1978. Grimsel-Oberrar pumped storage system[J]. *Water Power & Dam Construction*, 1978(2): 351-357.

WAN B, HARRIES K A, PETROU M F, 2002. Transfer length of strands in prestressed concrete piles[J]. *ACI Structural Journal*, 99(5): 577-585.

ZAREIFARD M R, FAHIMIFAR A, 2016. A simplified solution for stresses around lined pressure tunnels considering non-radial symmetrical seepage flow[J]. *KSCE Journal of Civil Engineering*, 20(7):2240-2654.

ZHANG Y, YAN Z, ZHU H, et al, 2018. Experimental study on the structural behaviors of jacking prestressed concrete cylinder pipe[J]. *Tunnelling and Underground Space Technology*, 73(3): 60-70.

Chapter 3
Mechanical characteristics of the MUAA lining under internal water loading

3.1 Introduction

Engineering practices show that prestressing linings with annular anchors offers the advantages of low prestress losses and uniformly distributed inward pressures, while also revealing a complex stress and strain mechanism between the annular anchor and the lining concrete, and resulting in local cracking of the lining near the block-outs during the operation of Xiaolangdi Desilting Tunnel (Cao et al, 2016; Lv et al, 2009). Clearly, the annular anchors of MUAA lining show a quite different force transfer mode, compared with that of a conventional prestressed anchor structure. The mechanical concept of the conventional prestressed annular anchors at the anchoring and tensioning ends is clear, that is, squeezing action between the two ends generates a compressive stress in concrete (Fahimifar and Soroush, 2005; Showkati et al, 2016). The circumferential tension of the anchor is converted to the radial load on the interface between the anchor and the lining through the "hoop effect", which is how a compressive prestress is generated in concrete. However, the unique winding pattern of the annular anchor in the MUAA lining results in an intersection of the anchors and the research progress in conventional prestressed anchor structures cannot be directly applied to MUAA lining. Therefore, further in-depth studies should be carried out for determining mechanical properties of the MUAA lining and establishing the corresponding analysis method, although several researches have been conducted on the stress characteristics and structural optimisation of prestressed linings with annular anchors (Zarghamee et al, 1993; Nishikawa, 2003; Nagamoto et al, 2008; Kang et al, 2014; Lee et al, 2015).

In order to address the technical problem mentioned above, a large-scale in-situ loading test of MUAA lining was carried out for the Songhua Water Transfer Project under construction and the corresponding theoretical analysis conducted. The variation in the tension of the annular anchors and the force transfer mechanism within the prestressing lining during the high internal water loading process were studied. An analytic solution to the stress state of the structure of the MUAA lining was proposed and compared with and verified by the in-situ loading test measurements.

3.2 In-situ test

3.2.1 Project background

The in-situ test was carried out at Songhua River Water Diversion Works, which is a large-scale long-distance water transfer project for solving the water supply problem in north-eastern China. The construction started in 2016 and is expected to be completed in 2020. The water delivery line is as long as 263.45 km. Therefore, pressure tunnels are used to improve the efficiency of long-distance water delivery, even though the overburden depth of the tunnel at some sections is quite low, as shown in Fig. 3.1. The surrounding rock is heavily weathered tufa of poor quality. The rock mass parameters are listed in Table 3.1. The minimum overburden depth of the tunnel is 11.8 m, the initial field stress is only about 0.28 MPa, and the maximum internal water pressure is 0.60 MPa. Under the action of the internal water pressure, the low field stress cannot meet the minimum principal stress criterion and the anti-lifting criterion with the hydraulic pressure tunnel lining design (Pachoud and Schleiss, 2016; Simanjuntak et al, 2016). As conventional reinforced concrete linings are vulnerable to large deformations and lining cracks, MUAA lining is used for resisting high internal water pressures. In order to investigate the mechanical properties of the lining under internal water pressures and ensure the structural safety, a large-scale in-situ loading test has been performed on a 3 m long tunnel with a typical section and MUAA lining.

Table 3.1 Parameters of the pressure tunnel, surrounding rock and MUAA lining for the in-situ testing

Item	Surrounding rock				
	Elasticity modulus	Poisson's ratio	Angle of internal friction	Cohesion	Initial stress
Symbol/Unit	E_r/GPa	μ_r	φ/(°)	c/MPa	q_0/MPa
Value	0.3	0.38	23	0.05	0.23
Item	MUAA lining			Tunnel	
	Elasticity modulus	Poisson's ratio	Thickness	Inner radius	Internal pressure
Symbol/Unit	E_c/GPa	μ_c	d/m	r/m	P_0/MPa
Value	35.01	0.20	0.45	3.45	0.60

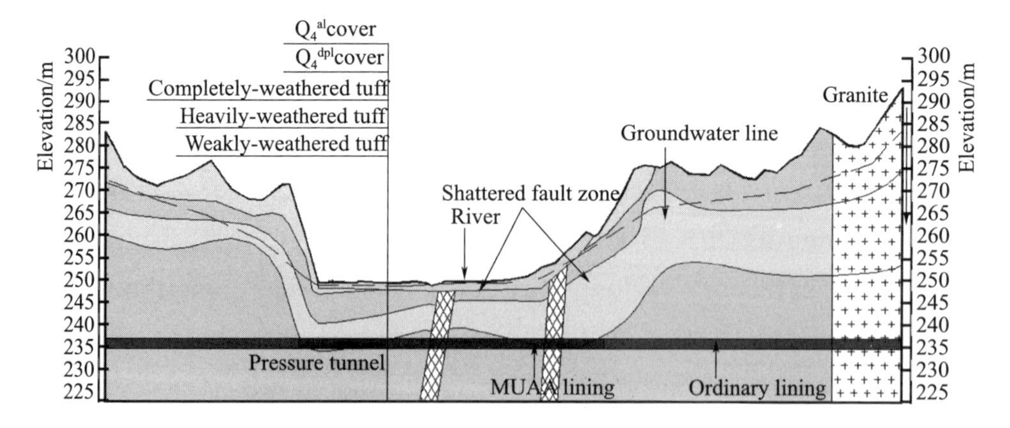

Fig. 3.1 Schematic diagram of the geological conditions of a pressure tunnel and the engineering section

3.2.2 Structure of MUAA lining

Fig. 3.2 shows the test plan for the MUAA lining. The anchorage block-outs are arranged crosswise on the inner side of the lower part of the lining. Anchor holes are set on both the anchoring end and the tensioning end of the free anchor head in the groove. Annular anchors are extended from the anchoring end to the outside of the lining, then tied to the outer conventional steel bars, which are encircled by the outer circumference at 720°, and finally fixed at the tensioning end (the typical one-layer and two-hoop type). The symmetrical structure of the lining with unbonded annular anchors ensures uniform distribution of the prestressed load.

The class of the lining concrete is C40, and the elasticity modulus of the reinforced concrete is 35 GPa. One bundle of prestressed annular anchors is composed of four steel strands with a spacing of 500 mm in the axial direction. For a single annular anchor, the designed prestress is 195.2 kN, the area of cross-section 140 mm^2, and the standard tensile strength 1860 MPa.

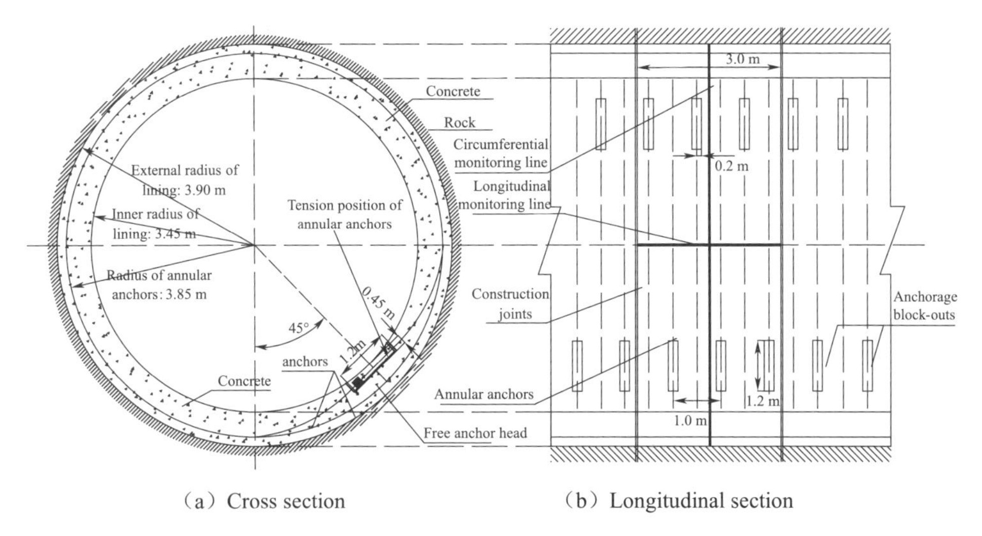

(a) Cross section (b) Longitudinal section

Fig. 3.2 Structural design drawings of MUAA lining for in-situ testing

3.2.3 Internal water load design

To realise a uniform, precise and stable internal water load on the MUAA lining, an internal water loading system with an annular flat jack was developed, as shown in Fig. 3.3. It mainly consists of a circular flat jack, reaction support, and pressurising device. The circular flat jack is composed of several arc-shaped flat jacks, and is connected by threaded hydraulic steel pipes to distribute the water pressure. The arc-shaped flat jack is a thin hollow pressure capsule which is welded by a thin steel plate; it is capable of injecting water via a high-pressure water pump to produce radial deformation, so that the pressure acts directly on the MUAA lining and the reaction support, thereby achieving internal water loading.

To facilitate the installation, a certain circumferential and longitudinal spacing is set between the arc-shaped flat jacks. As a result, the actual water pressure acting on the lining is slightly smaller than the designed value. Therefore, an internal water pressure compensation coefficient α is introduced:

$$P=\alpha P_0 \tag{3.1}$$

where: P_0 is the designed water pressure (MPa); P is the water injection pressure (MPa); α is the longitudinal and circumferential spacing parameter, which can be obtained by numerical simulation or from analytic solutions (Pi et al, 2018). According to the dimensions of the geometry, the α value in the test was 1.06.

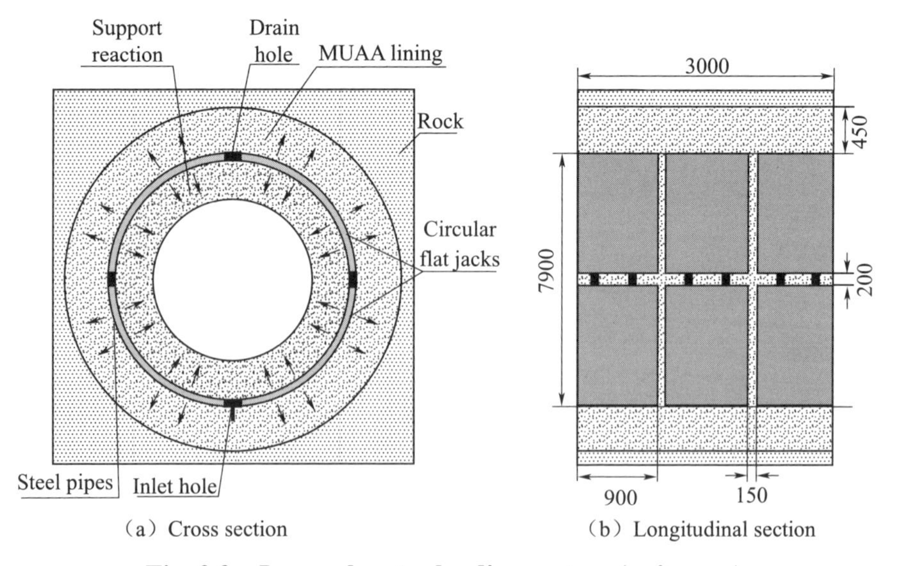

（a）Cross section （b）Longitudinal section

Fig. 3.3 Internal water loading system (unit: mm)

3.2.4 Test monitoring

As shown in Fig. 3.2, circumferential and longitudinal monitoring sections are set inside the MUAA lining in the in-situ test. The longitudinal section is used to monitor the lining stress state, while the circumferential sections (Fig. 3.4) are used to monitor the following:

(1) Lining prestress. Fibre grating sensors are embedded for measuring the circumferential prestress (no. FGS1-FGS16) and the longitudinal prestress (no. FGS17-FGS22) of concrete.

(2) Tension along the annular anchors. Magnetic flux sensors (no. C1-C7) are embedded for monitoring the tension along the annular anchors.

(3) Tension at the end of the annular anchors. A dynamometer of the annular anchor (no. N1) is placed at the anchor free head for monitoring the tension at the end of the annular anchors.

(4) Interaction force between the surrounding rock and the lining. Soil pressure gauges (no. T1-T5) are embedded in the areas between the surrounding rock and the crown, side wall and inverted arch of the lining for monitoring the contact stress.

(5) Surrounding rock deformation. Multi-point displacement meters are installed in the surrounding rock at the tunnel vault (no. X1) and side wall (no. X2) at depths of 3, 10 and 20 m.

(6) Contact mode between the surrounding rock and the lining. Joint meters (no. J1 and J2) are embedded in the area between the surrounding rock and the lining at the vault, side wall and so on for monitoring joint closure and opening width.

In addition, a pressure gauge is used to measure the values of the internal water load.

3.2.5 Internal water loading program

The internal water loads were set to 0.1, 0.2,···, 0.7 MPa in seven steps. The loading time was 130 min, except for the last step, for which it was 260 min. Since no relevant stability requirements were specified for internal

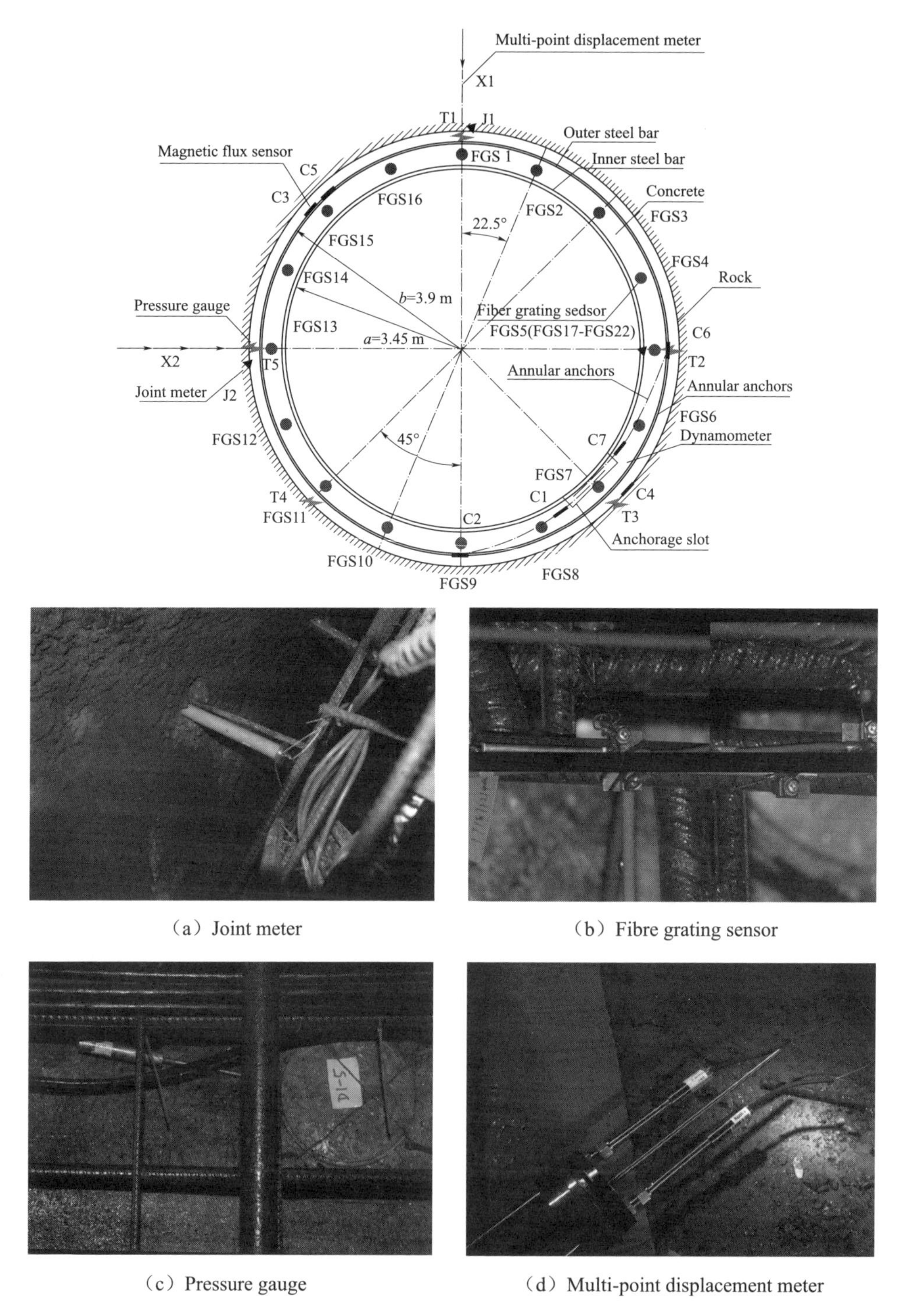

Fig. 3.4(1) Program plan and the equipment used for monitoring

(e) Magnetic flux sensor

(f) Dynamometer

Fig. 3.4(2) Program plan and the equipment used for monitoring

water loading for this type of in-situ tests in the pressure tunnel, the value of 0.05 με/min for the circumferential deformation of the lining per unit time was used as the control criterion for the in-situ test. 0.05 με represents about 1/2000th of the circumferential deformation of the lining during loading at the last step. In order to ensure the long-term stability of the MUAA lining and the surrounding rock, a pressure loading servo device composed of an ABB variable frequency pump and a multi-step pressure tank was developed for water injection. During loading, the pressure value on the control panel was set in advance. The loading accuracy is 0.02 MPa. The servo pressurising device would automatically apply the pressure. During the on-site loading process, the structural force was regarded as being stable if the measurements of the pressure gauges on the same level were almost identical, which indicated that the internal water loading achieved the expected effect.

3.3 Loads on the surrounding rock and the MUAA lining

The interaction between the surrounding rock and the lining has always been the fundamental issue in tunnelling and underground engineering. Tunnel excavation and related supports are generally considered to be the unloading process of the rock mass (Mezger et al, 2013) i.e., the external load (surrounding rock pressure, external water pressure, etc.) is transmitted from the surrounding rock to the lining. Once the internal water pressure of the tunnel exceeds the external force, the load (internal water pressure) will be transferred from the lining to the surrounding rock, which is a typical loading process for the rock mass.

In the in-situ test, when the prestressed MUAA lining was under an internal water load, the contact pressure curve for the surrounding rock and lining was measured; it is shown in Fig. 3.5. The contact stress gradually increases along with the increase in the load level. The internal water load is transferred from the inner side of the lining to the surface of the surrounding rock and deeper. When the load is 0.1-0.3 MPa, the contact stress essentially becomes stable after 60 min. When the load exceeds 0.6 MPa, the contact stress starts to decrease slightly after 130 min. Therefore, the changes in the contact stress show an apparent hysteresis effect, when compared with the internal water pressure.

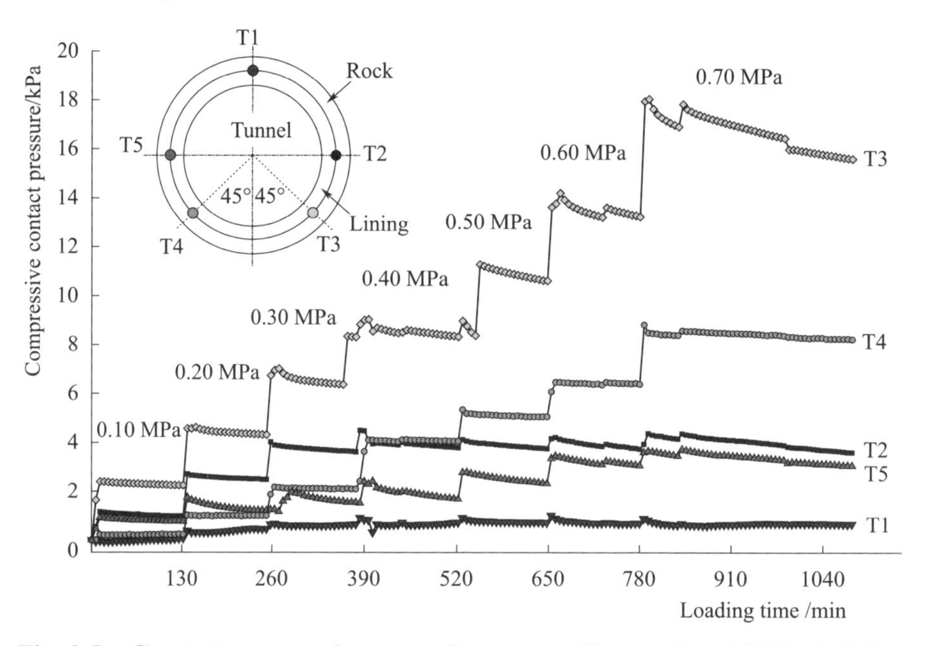

Fig. 3.5 Contact pressure between the surrounding rock and MUAA lining

The results of the spatial distribution of the contact pressure between the surrounding rock and the MUAA lining (Fig. 3.6) show that the contact pressure at the crown is much smaller than that at the inverted arch and that the maximum contact stress of 15.59 kPa appears near the inverted arch of the tunnel. The contact pressure between the surrounding rock and the lining i.e., the fraction of the internal water pressure which is transferred to the surrounding rock surface can be regarded as the internal water load shared with the surrounding rock. Therefore, when the internal water pressure is 0.7 MPa, the internal water loads shared by the surrounding rocks on the arch crown, left sidewall, right sidewall, left inverted arch and right inverted arch are only 0.10% (0.70 kPa/0.7 MPa), 0.44% (3.06 kPa/0.7 MPa), 0.51% (3.60 kPa/0.7 MPa), 1.17% (8.23 kPa/ 0.7 MPa) and 2.27% (15.59 kPa/0.7 MPa), respectively. Therefore, the bearing capacity of the surrounding rock is negligible. On the other hand, the MUAA lining shares more than 97% of the internal water load, and therefore, the bearing capacity can be regarded as being independent of the surrounding rock conditions.

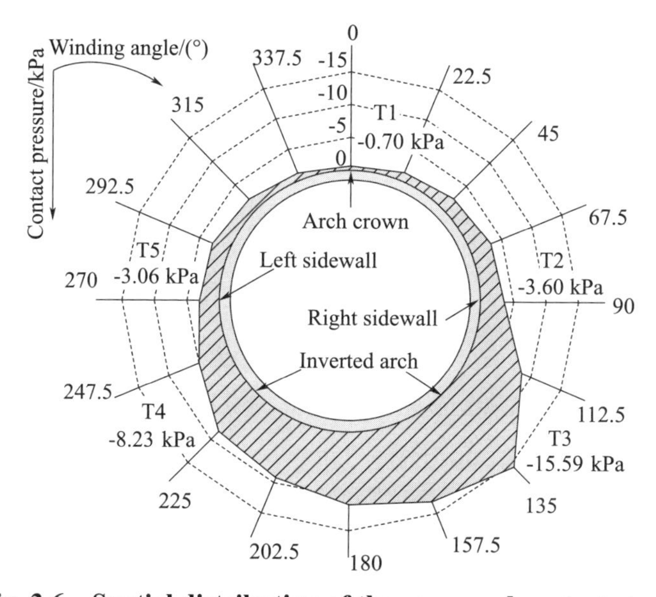

Fig. 3.6 Spatial distribution of the measured contact stress

The measurement results of the joint meter (Fig. 3.7) indicate that the MUAA lining has shrunk owing to the tension of the annular anchors before the internal water pressure is applied, therefore, the middle and upper parts of the lining will be separated from the surrounding rock (Kang et al, 2014; Cao et al, 2016).

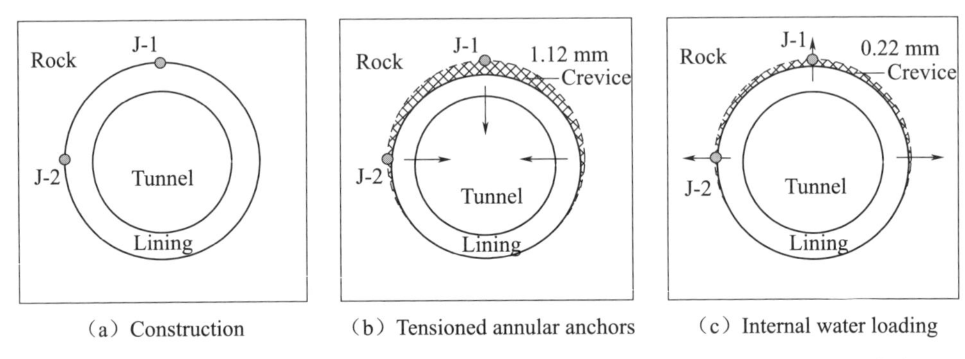

(a) Construction (b) Tensioned annular anchors (c) Internal water loading

Fig. 3.7 Change of crevice between the surrounding rock and MUAA lining

As shown in Fig. 3.7, there is a gap of 1.12 mm at the arch crown (J1). In the in-situ test, consolidation grouting was not carried out for the space behind the lining, therefore, the elastic deformation of the lining caused by the internal water loading could be absorbed by the space between the lining and the surrounding rock. According to the requirements of structural design, the equivalent radial prestress load should be larger than the internal water pressure. A 0.22 mm gap was still left between the lining and the surrounding rock at the arch crown after the internal water loading. The MUAA lining was designed to prevent the transfer of the internal water load to the surrounding rock. The measurements of the in-situ test proved that the MUAA lining can bear loads independent of the surrounding rock conditions and can be used as the pressure tunnel support even under very poor surrounding rock conditions. Therefore, MUAA lining is applicable to large-diameter pressure tunnels with thin overburdens, poor geological conditions, and high internal water pressures.

3.4 Tension of annular anchors

3.4.1 Distribution of tension along annular anchors

The tension of unbonded annular anchors is the cause of the prestress generated in a lining. Therefore, the overall uniformity of the lining prestress is determined directly by the distribution of tension along the annular anchors. The non-linear distribution of the tension of the annular anchors is mainly affected by the friction between the tendon and the duct, the non-circular arrangement in the vicinity of the anchorage block-out, and the loss of stress along the annular anchors. To measure the tension of the annular anchors at all the key parts of the lining structure during the stressing process, magnetic flux sensors were embedded.

The field monitoring data show that annular and conventional anchors differ significantly in their mechanical properties. For the conventional anchors, the maximum tension occurs at the stressing end and the minimum tension at the fixed end, yet, the in-situ test shows that the annular anchors exhibit similar tension values at the stressing and fixed ends. Therefore, the prestress loss from the centre to the fixed end can be significantly reduced by using unbonded annular anchors in the in-situ test. Therefore, conclusions can be drawn that the use of double-hoop winding mode can ensure an approximately uniform distribution of the prestress. The average value of the tension along double-hoop annular anchors (T) can be formulated as

$$T=\beta T_0 \tag{3.2}$$

where: T_0 denotes the designed prestress for the annular anchors (kN); β denotes the average loss factor, the value of which is assigned as 0.81 in this test.

The average tension curve of the annular anchors obtained during the internal water loading process is shown in Fig. 3.8. The average tension of the annular anchors is 157.7 kN before loading. The tension of the annular anchors increases gradually along with the increase in the internal water pressure. The tension of the annular anchors is 165.0 kN when the internal water load reaches the designed value of 0.6 MPa. Based on the average tension curve presented in Fig. 3.8, the Curve Fit Method was used for establishing the relationship between the increase in the tension of the annular anchor (ΔT, kN) and the internal water pressure (p_0, MPa):

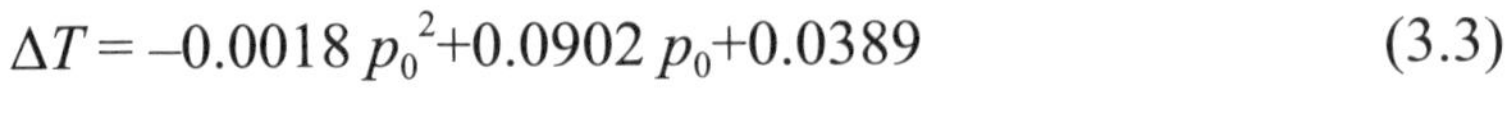

$$\Delta T = -0.0018\, p_0^2 + 0.0902\, p_0 + 0.0389 \tag{3.3}$$

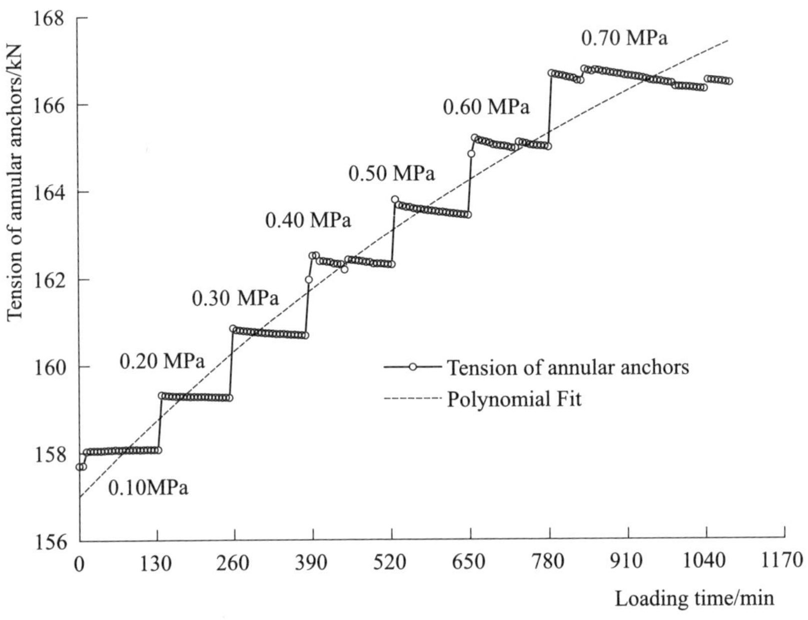

Fig. 3.8 Average tension of annular anchors during the loading process

In the in-situ test, seven ϕ5 mm, 1860 MPa high-strength steel strands were used, which is quite common in practice. Under the condition of poor quality of the surrounding rock, it can be considered that the MUAA lining mainly bears the load. When the bearing capacity of the surrounding rock is negligible, it is reasonable to apply Eq. (3.3) to other project sites. However, under the condition of combined bearing of the load by MUAA lining and the surrounding rock, the tension of unbonded annular anchors is affected by the surrounding rock, and the results calculated based on Eq. (3.3) are inaccurate.

3.4.2 Equivalent load of tension of annular anchors

The essential difference between the MUAA lining and a conventional prestressed structure is the action mode of the prestressed load. When analytic methods are used to analyse the mechanical properties of MUAA lining, the tension of the annular anchors should be converted into an inward surface load which is perpendicular to the tunnel wall. Therefore, the tension of the prestressed annular anchors should be converted into a load which is equivalent to the normal prestress, which is perpendicular to the interface between the annular anchors and concrete.

The differential element in Fig. 3.9 shows the load transfer between the prestressed annular anchors and the lining concrete. The tensions at the left and right ends of the annular anchors are T and T+dT, respectively, and the vertical loading width is B. It is assumed that the concrete stress is uniformly distributed on the differential element, the equivalent load of normal prestress is P_{p} and the angle between the centres of the differential element is dθ. The normal mechanical equilibrium condition is

$$\left(2T+\mathrm{d}T\right)\sin\frac{\mathrm{d}\theta}{2}-R\mathrm{d}\theta BP_{\mathrm{p}}=0 \tag{3.4}$$

where: R denotes the internal radius of the annular anchor section.

According to the design of MUAA lining, a polyethylene sleeve is placed between the annular anchors and concrete and filled with grease. In general, the friction coefficient for such a contact surface is less than 0.03 owing to curvature of the profile. To simplify the calculation, the friction was neglected (F=0), thus, dT=0. If dθ is small enough, $\sin\frac{\mathrm{d}\theta}{2}$ is expanded by using Taylor series and the higher-order terms are omitted, which then yields $\sin\frac{\mathrm{d}\theta}{2}=\frac{\mathrm{d}\theta}{2}$.By simplifying Eq. (3.4), the equivalent prestressed load of a single annular anchor can be formulated as

$$P_{\mathrm{p}}=\frac{T}{RB} \tag{3.5}$$

Therefore, the equivalent prestressed load under internal water loading can be obtained as

$$P_{\mathrm{p}} = \frac{T + \Delta T}{RB} \tag{3.6}$$

where: ΔT denotes the increase in the tension of a single annular anchor.

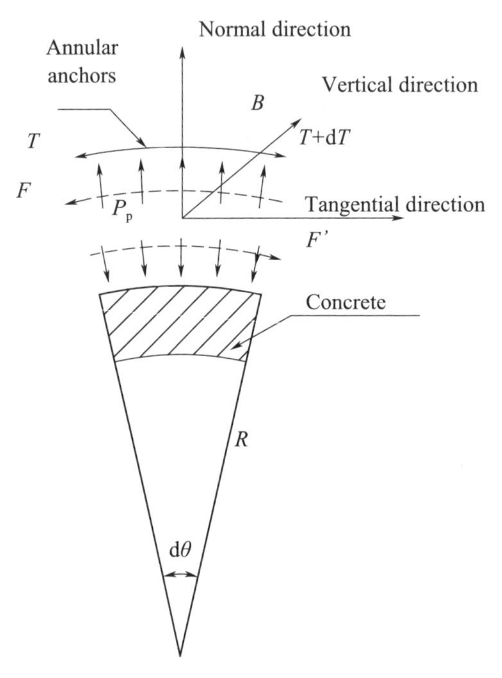

Fig. 3.9 Illustration of Load transfer between the prestressed annular anchors and the lining concrete

3.5 Prestress redistribution of MUAA lining

3.5.1 Stress analysis of MUAA lining

As mentioned above, the load-bearing capacity of the surrounding rock under the action of internal water pressure is negligible. The MUAA lining mainly bears the internal water load and the external prestressed load (P_{p}), and is designed to be a crack-free structure in principle. Thus, the water pressure P_0 acting on the inner surface can be considered as a surface force. Fig. 3.10 shows that the structure of MUAA lining is a typical thick-walled cylinder under uniform internal and external pressures. In the case of axial symmetry, the stress and strain generated in the thick wall are also axisymmetric. Consequently, the shear stress and the shear strain are both zero. Therefore, according to the principle of elastic mechanics, the Lamé equation can be obtained for the thick-walled cylinder under internal and external pressures (Sulem et al, 1987; Zareifard and Fahimifar, 2016). For the tunnel, the axial stress of lining (σ_a, MPa) is very small, so it can be ignored (Sulem et al, 1987). The radial stress (σ_r, MPa) and the circumferential stress (σ_θ, MPa) inside the lining are given by

$$\left.\begin{aligned}\sigma_r &= \frac{P_0 a^2 - P_\text{p} b^2}{b^2 - a^2} - \frac{\left(P_0 - P_\text{p}\right) a^2 b^2}{\left(b^2 - a^2\right) r^2} \\ \sigma_\theta &= \frac{P_0 a^2 - P_\text{p} b^2}{b^2 - a^2} + \frac{\left(P_0 - P_\text{p}\right) a^2 b^2}{\left(b^2 - a^2\right) r^2}\end{aligned}\right\} \tag{3.7}$$

where: a is the inner radius of the lining; b is the outer radius of the lining; P_p is the equivalent prestress load under internal water loading; and P_0 is the internal water pressure.

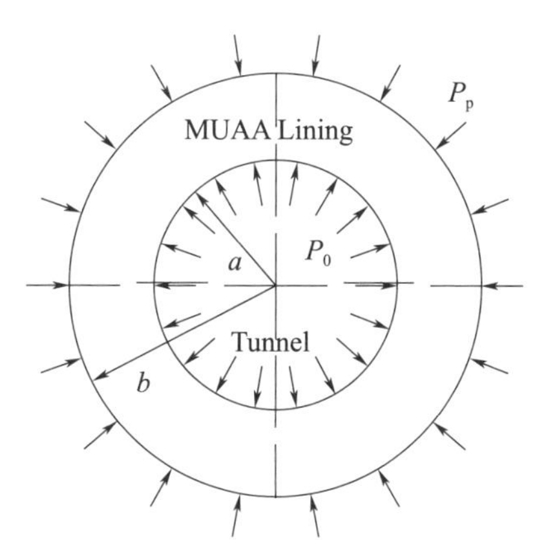

Fig. 3.10 Mathematical model of MUAA lining

Then, the circumferential stress σ_θ in Eq. (3.7) is integrated with respect to the centroid of the section:

$$M = \int_b^a \sigma_\theta \left(r - \frac{a+b}{2} \right) \mathrm{d}r \tag{3.8}$$

where: r denotes the distance between the calculated point and the centre of the tunnel.

By solving Eq. (3.8), the sectional bending moment of the structure can be obtained:

$$M = \frac{\left(P_0 - P_{\mathrm{p}}\right) a^2 b^2}{b^2 - a^2} \ln \frac{b}{a} - \frac{1}{2}\left(P_0 - P_{\mathrm{p}}\right) ab \tag{3.9}$$

Three-dimensional stress (axial stress, circumferential stress, and radial stress) state of the thick-walled cylinder can be calculated according to Lame equation (Sadd, 2009). Thick-walled cylinder theory and Lame equation take into consideration the three-dimensional stress state of annular structure for both thick-walled and thin-walled cylinders. For thin-walled cylinders, the solution can be further simplified based on the Membrane theory, in which the stress is considered uniformly distributed along the wall thickness and the radial stress of the vertical and cylinder wall are ignored (Sadd, 2009). This solution is an approximate method. In this study, the thick-walled cylinder theory is used to calculate the structural stress of MUAA lining. So, the obtained formulas can also be applied to thin-walled MUAA lining.

3.5.2 Measurement results of prestress redistribution

Fig. 3.11 shows the stress redistribution curves of MUAA lining during the internal water loading process. The measurements (black lines) show that the overall prestress of the lining with annular anchors is uniform. The analytical results of the prestress obtained using Eq. (3.7) (green lines) are in good agreement with those obtained from the in-situ test, especially for 0.2 MPa and 0.3 MPa internal water pressures. Hence, the formula for calculating the prestress [Eq. (3.6)] which was established by the equivalent load method proposed in this paper can well describe the stress state of the lining structure.

Before the internal water loading, the average initial prestress of the lining is 6.11 MPa [Fig. 3.11(a)]. When it is equal to the design load [0.6 MPa, Fig. 3.11(g)], the overall prestress of the structure is reduced by 3.19 MPa, on average, and a prestress of 0.36-4.16 MPa can still be maintained; the MUAA lining displays favourable safety. Under overloaded internal water pressure [0.7 MPa, Fig. 3.11(h)], the lining shows a tensile stress of 0.28 MPa on the side (FGS6), which is still smaller than the allowable tensile stress of C40 concrete, and the lining is still in an elastic state without any cracks.

As shown in Fig. 3.11, for all the internal water pressure levels, the points FGS5 (90°, at the right wall) and FGS13 (270°, at the left wall) reveal larger measured prestresses than other points both before and after loading. It is mainly because the anchorage block-outs are arranged at the angles of 122.5°-135° and 225°-247.5° (see Fig. 3.2). This arrangement leads to non-uniform local stress and stress concentration. In addition to the normal prestress, the nearby concrete will also generate an additional stress. The superimposition of the original stress and the additional stress will increase the local stress. Meanwhile, the prestress behind the anchorage block-outs (112.5° and 247.5°) will be reduced and, consequently, a weak area will be developed in the MUAA lining. It is the main cause of the tensile stress observed at the side of the lining under the overloaded water pressure (FGS6). As a result, the area behind the block-out is susceptible to damage under high internal water pressures.

The results of prestress redistribution of MUAA lining also show that the areas with larger prestresses before loading tend to exhibit greater reductions after loading, but with similar reduction ratios. For example, the initial prestress at the centre of the lining [FGS13 in Fig. 3.11(a)] is 7.87 MPa and the prestress is reduced to 4.49 MPa after loading [Fig. 3.11(g)]. The corresponding reduction ratio is 57.05%. The initial prestress at the bottom of the lining (FGS9) is 4.14 MPa and the prestress is reduced to 2.73 MPa after loading [Fig. 3.11(g)]. The corresponding reduction ratio is 65.94%. Although the prestress values differed greatly, the reduction ratios are similar.

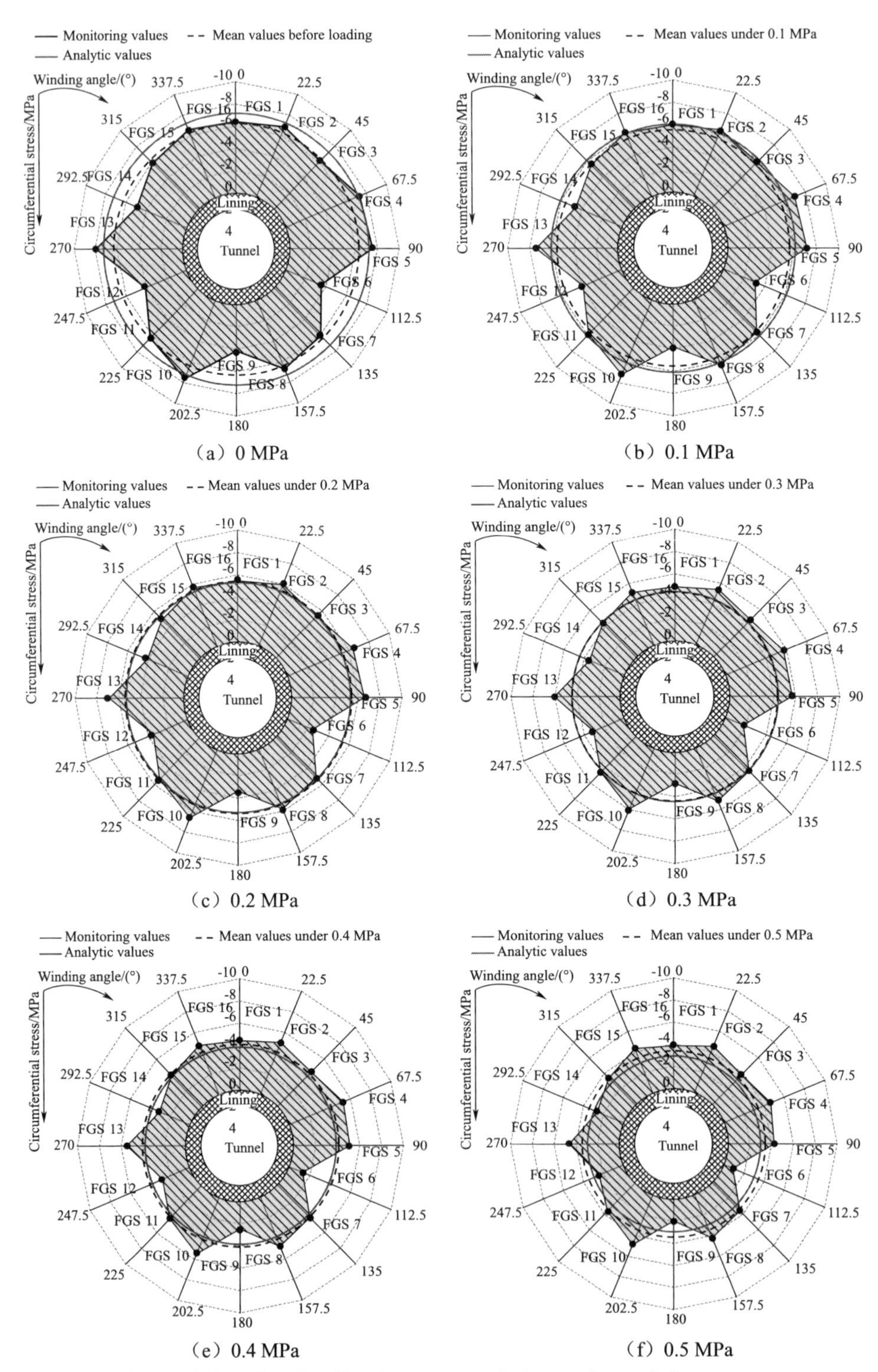

Fig. 3.11(1) Redistribution maps of circumferential prestress

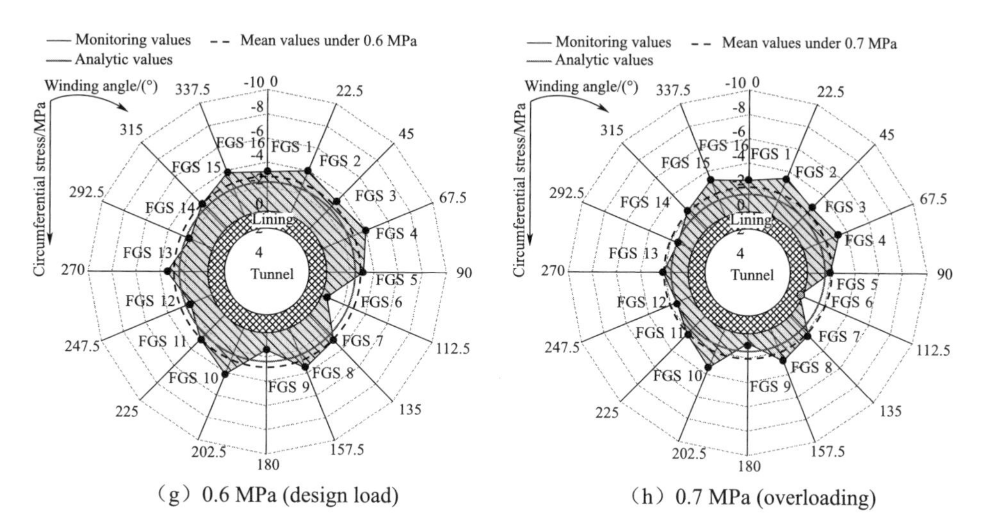

(g) 0.6 MPa (design load)　　(h) 0.7 MPa (overloading)

Fig. 3.11(2)　Redistribution maps of circumferential prestress

As shown in Fig. 3.12, the prestresses along the vertical lining are essentially identical. Therefore, it is reasonable to assume that the interaction between the tunnel and the lining is a plane stress-strain problem (see Fig. 3.10).

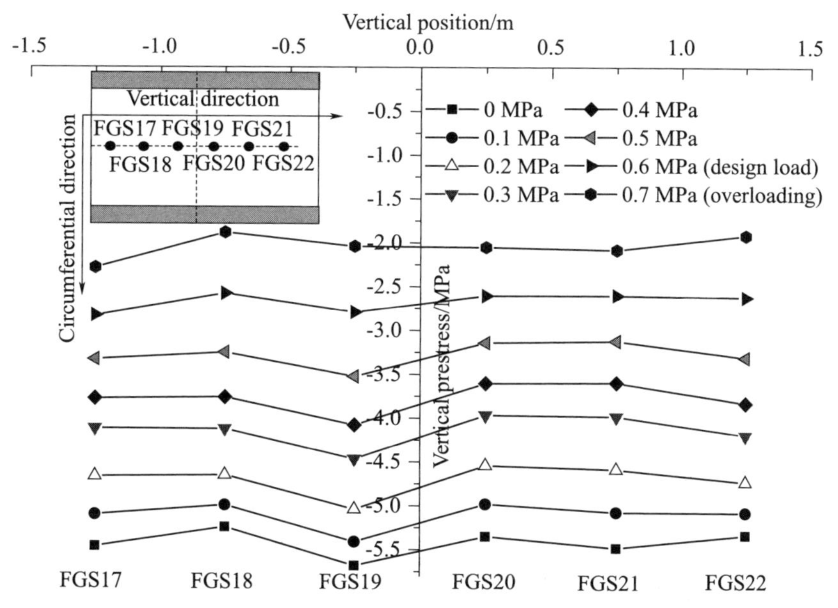

Fig. 3.12　Redistribution curves of vertical prestress

The distributions of the prestress of MUAA lining measured during the loading process are presented in Table 3.2. As the internal water load increases, the standard deviation of the prestress value reduces gradually from 1.32 to 1.01. It indicates that the values of the lining prestress tend to be consistent with the internal water loading and that the spatial distribution of the prestress is more uniform, which can effectively slow down the tendency of the lining to crack in the area with small prestresses.

Table 3.2 Statistical characteristics of the prestress distribution

Internal water load/ MPa	Max/MPa	Min/MPa	Mean/MPa	Standard deviation
0	-3.54	-7.87	-6.11	1.32
0.1	-3.05	-7.44	-5.73	1.31
0.2	-2.43	-6.86	-5.21	1.28
0.3	-1.74	-6.04	-4.57	1.22
0.4	-1.28	-5.47	-4.15	1.19
0.5	-0.87	-4.86	-3.60	1.11
0.6 (Design value)	-0.36	-4.16	-2.92	1.03
0.7 (Overload)	0.28 (Tension)	-3.42	-2.19	1.01

3.6 Discussion and conclusions

Concrete linings actively prestressed by posttensioning unbonded annular anchors provide a sound solution to problems of pressure tunnels such as an inadequate overburden depth or adverse surrounding rock conditions. A new type of prestressing structure, the MUAA lining, is proposed in this research. To further explore the mechanism of mutual action among the annular anchor, concrete lining and surrounding rock, a new large-scale in-situ loading test is proposed and carried out at a 6.8 m diameter water conveyance tunnel in Jilin, China, in which case the internal water pressure is applied by an annular flat jack. Variation in the tension of the annular anchor and the characteristics of prestress redistribution in the lining during posttensioning the tendon and application of the internal water load are revealed. Furthermore, analytic solutions to the lining stress and bending moment induced by annular anchors and internal water pressure are proposed, which are verified by the field test results.

The tension along the annular anchors displays a non-linear distribution and that the use of double-hoop winding mode can superpose the forces to overcome the difference in stress at each part of the lining and ensure an approximately uniform distribution of the overall prestress of the lining. The circumferential and longitudinal prestresses of the lining tend to be consistent during the internal water loading process, which will effectively slow down the tendency of the lining to crack in areas where the prestress is low.

An analytical method for converting the tension of annular anchors into an equivalent prestress load was proposed and verified based on the field test results. The analytic formulas of the lining stress, annular anchor force and surrounding rock load sharing were established and can be applied to evaluate other MUAA lining with similar boundary conditions.

MUAA lining exhibits high tensile strength and impermeability and can sustain loads independent of the surrounding rock conditions. It provides a new structure for support design of large-diameter pressure tunnels with thin overburden, poor geological conditions, and high internal water pressure.

References

CAO R L, WANG Y J, ZHAO Y F, et al, 2016. Study on numerical modeling method of the prestressed tunnel lining with unbonded curve anchored tendons[J].

Journal of China Institute of Water Resources and Hydropower Research, 14(6): 471-477.

FAHIMIFAR A, SOROUSH H, 2005. A theoretical approach for analysis of the interaction between grouted rock bolts and rock masses[J]. *Tunneling and Underground Space Technology*, 20(4): 333-343.

KANG J F, SUI C E, WANG X Z, 2014. Structure optimization of the prestressed tunnel lining with unbonded circular anchored tendons[J]. *Journal of Hydraulic Engineering*, 45(1): 103-108.

LEE Y, LEE E T, 2015. Analysis of prestressed concrete cylinder pipes with fiber reinforced polymer[J]. *KSCE Journal of Civil Engineering*, 19(3): 682-688.

LV L, YANG Y, TANG W, 2009. Treatment for oil and water penetration in anchorage channel of san discharge tunnel, Xiaolangdi key water control project[J]. *Chinese Building Waterproofing*, 2009(2): 36-38.

MEZGER F, ANAGNOSTOU G, ZIEGLER H J, 2013. The excavation-induced convergences in the Sedrun section of the Gotthard base tunnel[J]. *Tunnelling and Underground Space Technology*, 38(9): 447-463.

NAGAMOTO, TAKAYUKI, YONEDA. 2008. Large P & PC segment works: construction works for rainwater storage under Osaka International Airport[J]. *Underground and Tunnel*, 39(8): 581-589.

NISHIKAWA K, 2003. Development of a prestressed and precast concrete segmental lining[J]. *Tunnelling and Underground Space Technology,* 18(2): 243-251.

PACHOUD A J, SCHLEISS A J, 2016. Stresses and displacements in steel-lined pressure tunnels and shafts in anisotropic rock under Quasi-static internal water pressure[J]. *Rock Mechanics and Rock Engineering*, 49(4): 1263-1287.

PI J, WANG X G, CAO R L, et al, 2018. Innovative loading system for applying internal pressure to a test model of prestressed concrete lining in pressure tunnels[J]. *Journal of Engineering Research*, 6(2): 24-44.

SADD M H, 2009. *Elasticity, Theory, Applications, and Numerics*[M]. Oxford: Academic Press, Oxford, pp: 172–174.

SHOWKATI A, MAAREFVAND P, HASSANI H, 2016. An analytical solution for stresses induced by a post-tensioned anchor in rocks containing two perpendicular joint sets[J]. *Acta Geotechnica*, 11(2): 415-432.

SIMANJUNTAK T D Y F, MARENCE M, SCHLEISS A J, et al, 2016. The interplay of in situ stress ratio and transverse isotropy in the rock mass on prestressed concrete-lined pressure tunnels[J]. *Rock Mechanics and Rock Engineering*, 49(11): 4371-4392.

SULEM J, PANET M, GUENOT A, 1987. An analytical solution for time-dependent displacements in a circular tunnel[J]. *International Journal of Rock Mechanics and Mining Sciences & Geomechanics Abstracts*, 24(3): 155-164.

ZAREIFARD M R, FAHIMIFAR A, 2016. A simplified solution for stresses around lined pressure tunnels considering non-radial symmetrical seepage flow[J]. *KSCE Journal of Civil Engineering*, 20(7): 2240-2654.

ZARGHAMEE M S, OJDROVIC R P, DANA W R, 1993. Coating delamination by radial tension in prestressed concrete pipe. II: analysis[J]. *Journal of Structural Engineering*, 119 (9): 2720-2732.

Chapter 4
Numerical modeling method of the MUAA lining and verification

4.1 Introduction

It is a quite difficult task to design the MUAA lining, due to the complexity of the structure, the unclear mechanical mechanism of the interaction between the surrounding rock, the lining and the pressurized water (Wang et al, 2020). Because it is a new type of hydraulic structure, there is a lack of practical design specifications (Kang et al, 2014). The asymmetric distribution of structural mechanics model caused by friction loss in prestressed system makes it difficult to make stress and strain analysis directly by analytical method (Cao et al, 2016; Hajali et al, 2016). The most critical focuses include the constraints of surrounding rock on lining, the mechanical properties of unbonded annular anchor change, and the difficulty in analyzing the nonlinear prestress loss (Anderson et al, 2008; David et al, 2014). Therefore, we need to use numerical simulation to analysis the mechanical characteristics of the MUAA lining.

This study focuses on the structural mechanical properties and the corresponding numerical modeling mechanism of the MUAA linings, as well as the corresponding numerical modeling method and calculation method. Then, based on the finite difference software FLAC3D and engineering case, the mechanical characteristics of the MUAA linings are analyzed, and the calculation results are compared with the monitoring data to verify the correctness of the numerical modeling method.

4.2 Difficulties in numerical modeling

The composition of a typical prestressed reinforced concrete lining with single-layer double-hoop unbonded annular anchor is shown in Fig. 4.1. Three key problems in numerical modeling are analyzed in the following sections.

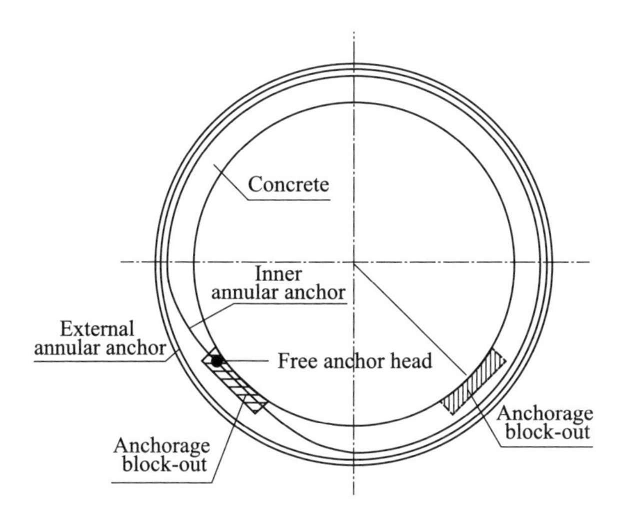

Fig. 4.1 A typical prestressed lining with single-layer double-hoop unbonded annular anchor

4.2.1 Constraint on lining by surrounding rock

Constraint on lining by surrounding rock is not invariable, and the contact relationship between the surrounding rock and the MUAA lining in pressure tunnels appears recurrent alternation of fitting and disconnecting during the construction period, operation period and maintenance period:

(1) During the construction period, the annular anchor is tensioned by the jack, and the lining is shrunk. The surrounding rock is separated from

the middle and upper lining, and with the increase of the stress, the gap between the surrounding rock and lining becomes bigger. So, the middle and lower surrounding rock only supports the concrete lining like a base.

(2) After backfilling grouting behind the lining, the surrounding rock wraps the lining, and the two are closely fitted.

(3) During the operation period, the internal water pressure makes the surrounding rock and lining squeeze each other and bear the internal water load together.

(4) During the maintenance period, the internal water pressure decreases, and there will be a small gap between the surrounding rock and the lining at the top of the tunnel.

Therefore, we need to establish a reasonable contact relationship according to the surrounding rocks' constraint on lining for numerical modeling. Not only to ensure that the maximum tensile stress on the surface of the lining and surrounding rock border exceeds the compressive strength between the two in order to cause automatically detachment and no longer transfer contact stress, but also to ensure that after re-fit, the surrounding rock and lining can jointly bear the internal water pressure transmitted by the lining inside.

4.2.2 Mechanical properties of unbonded annular anchor

The unbonded annular anchor is embedded in the smooth grease coated PE pipe, and is not integrated with the concrete casting. Therefore, the anchor and the surrounding concrete cannot meet the conditions of coordinated deformation. The structural elements (e.g., Link and Beam element in ANSYS (Kang et al, 2006; Tahmasebinia et al, 2012), Beam element in ABAQUS (Wang et al, 2013), Beam or Pile element in FLAC3D (Ghazavi et al, 2014), etc.) built in existing numerical software to simulate the mechanical properties of bonded annular anchor are not applicable to unbonded annular anchor. Unbonded post-tensioned prestressed annular anchors show different mechanical states in different engineering stages:

(1) During the construction period, after the annular anchor is tensioned, the annular tension inside the lining is transformed into radial stress and normal stress on the concrete interface. Due to the stress relaxation caused by the concrete shrinkage and creep, the pre-added stress will gradually decrease from a high value.

(2) During the operation period, after the internal water pressure is applied, the unbonded prestressed annular anchor has been fixed, and will bear the compressive stress transmitted by the lining concrete like the non-prestressed steel bar, thus increasing its tensile force.

(3) During the maintenance period, the increased tension will disappear as the internal water pressure decreases, and when the internal water pressure is applied, the prestress will increase.

Theoretically, we can use equivalent load method, temperature transfer stress method (Xiang, 2014), initial strain method and solid modeling tension method (Kang et al, 2006; Fahimifar & Soroush, 2005) to simulate the stress problem of curvilinear stress bars. The basic principle of equivalent load method is to convert the circumferential tension of anchor into the normal equivalent load and tangential equivalent load through the theoretical formula, acting on the lining concrete, so that the prestress can be generated. Since the formula can only calculate the constant initial prestress value, this method can hardly simulate the changing prestress value. Temperature transfer stress method and initial strain method both assume a connection between the annular anchor and the concrete model nodes, the forced position is converted into stress and applied to the lining through deformation coordination, therefore, these two methods can better simulate the bonded annular anchor. But the unbonded annular anchor does not conform to the deformation coordination, so they don’t work on it. The idea of solid modeling tension method is to directly establish a solid model of annular anchor, and establish an Interface (contact surface) property on the outside of the annular anchor, then tension the annular anchor to prestress the concrete. The mechanism of mechanical action of this method is consistent with the non-bonding system, and it can accurately simulate the prestressed curvilinear bars of the single-hoop wrapping method. For the double-hoop wrapping method, the annular anchor is difficult to be tensioned because of the cross problem of the prestressed curved bars. Therefore, for the numerical modeling of prestressed lining structures with unbonded curved anchors, none of the above methods is perfect, but they can be skillfully combined to form a solution.

4.2.3 Non-linear distribution of prestress loss

The prestress loss of annular anchor is an important factor that affects the overall prestress effect of lining concrete and relates to the value of prestress at the tensioning end. The prestress loss of annular anchor mainly includes: friction loss (σ_1), deflector tension loss (σ_2), annular anchor retraction loss (σ_3), annular anchor relaxation loss (σ_4), anchor stress loss caused by concrete creep (σ_5). σ_2-σ_5 can be realized by reducing the prestress value of the tensioning end, but σ_1 is distributed along the path and nonlinear. Therefore, it is necessary to calculate the distribution load of the prestress loss along the path during numerical modeling, and distribute it gradiently on the surface of the annular anchor.

4.3 Numerical simulation method of the MUAA lining

4.3.1 Principles of numerical modeling

(1) Constraint on the MUAA lining by surrounding rock. Various states of constraint on lining by surrounding rock can be simulated by establishing reasonable contact relationships, and the contact surface properties can be established by using Mohr-Coulomb shear strength criterion. Fig. 4.2 is the schematic diagram of the element principle of the constitutive model of the contact surface. The contact surfaces have properties such as sticking close to each other, sliding towards each other, and disconnecting from each

other, and the contact force is transmitted through nodes.

The state of the contact surface is determined according to the strength criterion. The tangential force $\tau_{s\max}$ required for the contact surface to make relative sliding of is:

$$\tau_{s\max} = c_{if} A + \tan\varphi_{if}\left(\sigma_n - PA\right) \tag{4.1}$$

Where: c_{if} is the cohesion of contact surface; φ_{if} is the friction angle of contact surface; A is the representative area of contact surface nodes (Fig. 4.2); σ_n is the substantive normal stress on the boundary surface; P is pore-water pressure.

The normal shear deformation leads to the increase of effective normal stress, and the normal force $\sigma_{n\max}$ required for the contact surface to disengage is:

$$\sigma_{n\max} = \sigma_n + \frac{\tau_s - \tau_{s\max}}{Ak_s} k_n \tan\gamma_{if} \tag{4.2}$$

Where: τ_s is the shear strength of the boundary surface entity; k_n is normal stiffness; k_s is shear stiffness; γ_{if} is the dilatancy angle of the contact surface.

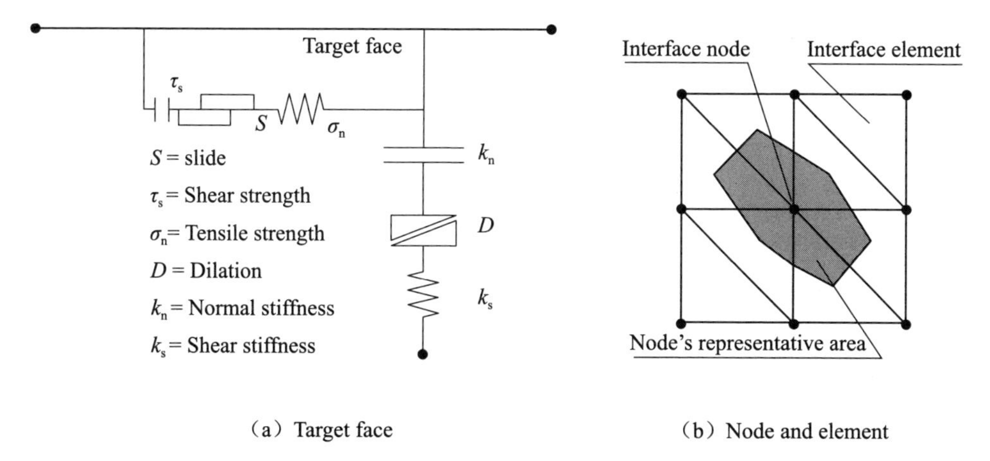

(a) Target face (b) Node and element

Fig. 4.2 Components of the bonded interface constitutive model

When the tangential force $|\tau_s|$ on the contact surface is less than $\tau_{s\max}$, the contact surface is in an elastic state without slip. When $|\tau_s| = \tau_{s\max}$, the contact surface is in a slip plastic state, and the shear force remains unchanged during the slip process. When the tensile stress on the contact surface exceeds the tensile strength σ_{nmax} of the contact surface, the contact surface is in a state of detachment, and the corresponding node no longer transfers tangential force and normal force. When the geometrical relationship of the elements on both sides of the contact surface is extruded into a close joint, the force transfer mechanism of the contact surface is restored.

(2) Mechanical properties of unbonded annular anchor. The joint method of equivalent load and solid modeling is used to simulate the force transmission process of unbonded prestressed annular anchor. According to the mechanical characteristics of unbonded prestressed anchor in concrete lining, the annular anchor stress is divided into constant prestressed component and variable non-prestressed component.

During numerical modeling, firstly, the Interface property is established on the interface between the annular anchor and the concrete to simulate the friction and slip relationship between the steel strand and the casing. Then, as shown in Fig. 4.3 (θ is the angle between the calculated section of the anchor and tensioning end, and α is the surrounding angle of the anchor), equal normal load and tangential load are used to simulate the unchanged prestress after annular anchor is fixed, while the solid model is used to simulate the changing non-prestress of annular anchor. Finally, the stress and strain states of annular anchor in different engineering stages are calculated through superposition of forces.

(3) Non-linear distribution of prestress loss. The geometric model used to calculate the friction loss of the unbonded annular anchor can be seen in Fig. 4.3(b), the outer annular of the anchor is a closed circle, and the inner annular includes large arcs, small arcs and straight lines. According to the geometric dimensions in the figure, the friction loss distribution coefficient is calculated according to Eq. (4.3) and Eq. (4.4). The distribution coefficient of geometric feature points is shown in Table 4.1. After the

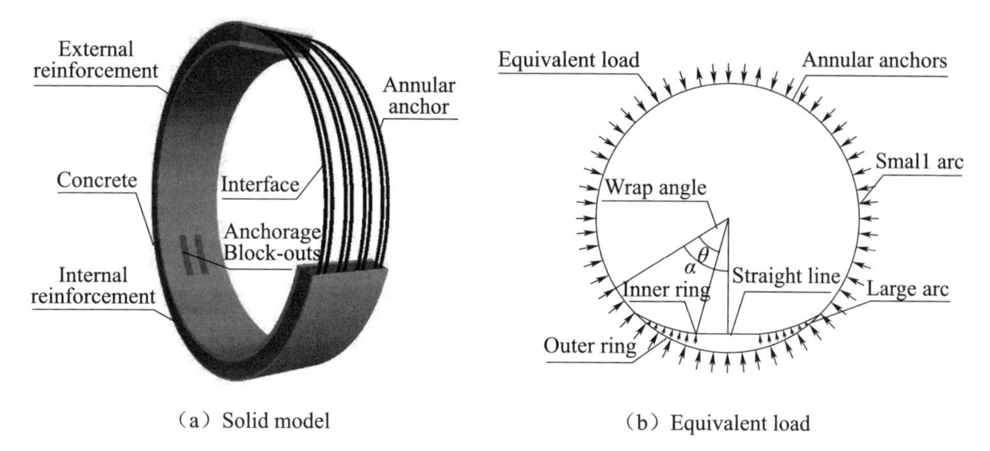

(a) Solid model (b) Equivalent load

Fig. 4.3 Superposition effect of equivalent load and solid model to simulate the annular anchor

prestress value is reduced by σ_2-σ_5, the continuous actual prestress value and the corresponding equivalent load value can be calculated by the friction loss distribution coefficient. Then, the nonlinear equivalent normal load and tangential load are applied to the nodes of the annular anchor model according to the gradient law by programming language.

$$\beta_1=\begin{cases} e^{-(kR\theta_1+\mu\theta_1)}-2L_f\left(\dfrac{\mu}{R}+k\right)\left(1-\dfrac{R\theta_1}{L_f}\right) & R\theta_1\leqslant L_f \\ e^{-(kR\theta_1+\mu\theta_1)} & R\theta_1>L_f \end{cases} \tag{4.3}$$

$$\beta_2=\begin{cases} e^{-(kR\theta_2+\mu\theta_2)}-2L_f\left(\dfrac{\mu}{R}+k\right)\left(1-\dfrac{R\theta_2}{L_f}\right) & R\theta_2\leqslant L_f \\ e^{-(kR\theta_2+\mu\theta_2)} & R\theta_2>L_f \end{cases} \tag{4.4}$$

Where: k and μ are the friction coefficient and deviation coefficient of the anchor, which are determined by the field friction test; β_1 and β_2 are the distribution coefficient of friction loss of the first and second annular anchor; θ_1 and θ_2 are the angle between the calculated section and tensioning end of the first and second annular anchors; L_f is the influence range of anchorage retraction.

Table 4.1 The distribution coefficient of geometric feature points friction loss of unbonded annular anchor

Wrapping angle	Angle		Distribution coefficient		Wrapping angle	Angle		Distribution coefficient	
α/(°)	θ_1/(°)	θ_2/(°)	β_1	β_2	α/(°)	θ_1/(°)	θ_2/(°)	β_1	β_2
0	-	350	-	0.770	225	215	125	0.852	0.651
10	0	340	0.855	0.765	270	260	80	0.824	0.630
45	35	305	0.880	0.745	315	305	35	0.797	0.609
90	80	260	0.915	0.720	350	340	0	0.776	0.594
135	125	215	0.911	0.697	360	350	-	0.770	-
180	170	170	0.881	0.974	-	215			

4.3.2 Numerical calculation scheme and modeling

(1) Calculation parameters. During the operation period, the maximum inner water head of the prestressed lining tunnel of the main line of a project is 66 m, and the minimum overburden thickness above the vault of local tunnel sections is very low. The prestressed lining test section uses C40 concrete, with an inner radius of 3.45 m and a wall thickness of 0.45 m. The elasticity modulus of concrete is set as E_c =3.25×10^4 MPa, Poisson's ratio μ=0.167, and the design value of axial tensile strength f_c =19.5 MPa; the design value of axial tensile strength f_t=1.80 MPa; concrete dead weight (lining dead weight) is set as 2400 kg/m^3, which means γ_c=23.52 kN/m^3.

The standard strength value of prestressed anchor is f_{ptk}=1860 MPa, and the design strength value is f_{py}=1260 MPa. Each unbonded bar is composed of 7ϕ5 high-strength low-relaxation steel strand, and the nominal diameter is D_n=15.2 mm; cross-sectional area S_n=140 mm^2; mass per meter G_n=1.101 kg.

In numerical calculation, considering the complexity of the model and the large number of elements, if the external water load is applied according to the seepage force, the convergence rate of the model calculation will be greatly reduced, and the external water pressure is small (the buried depth is low). Therefore, the external water pressure is loaded by equivalent surface force. The physical and mechanical parameters of the model are calculated according to the results of indoor and field tests.

The mechanical parameters of the contact surface between the surrounding rock and the lining and between the annular anchor and the concrete are shown in Table 4.2.

(2) Geometric dimensions. In the calculation and analysis of annular anchor lining, the distance among numerical model annular anchors is 50 cm, the thickness of lining is 45 cm, and the size of anchorage block-out is 120 cm×20 cm×20 cm. Considering the influence of boundary effect on the calculation results, the value range of surrounding rock is more than 2.5 times of the tunnel diameter.

Table 4.2 Mechanical properties of contact surface

Parameters	Normal stiffness k_n/(GPa/mm)	Shear stiffness k_s/(GPa/mm)	Cohesion c_{if} /MPa	Friction angle φ_{if}/(°)	Dilatancy angle γ_{if} /(°)
Between surrounding rock and lining	2.63	1.82	0.3	29	25
Between anchor and concrete	32.9	25.5	0	0	0

(3) Numerical model. In numerical modeling, the hexahedral grid solid elements subject to linear elasticity criterion are used for lining, the hexahedral grid solid elements subject to Mohr-Coulomb elastoplastic criterion are used for surrounding rock, the "solid + equivalent load + contact surface" model is used for prestressed annular anchor, and the Shell model is used for non-prestressed tendon. The Interface property which is subject to Mohr's yield criterion is set between the lining and the surrounding rock to simulate the states of them clinging to each other, sliding towards each other and disconnecting from each other (Fig. 4.4).

The three-dimensional numerical model of annular anchor lining is shown in Fig. 4.5 and Fig. 4.6.

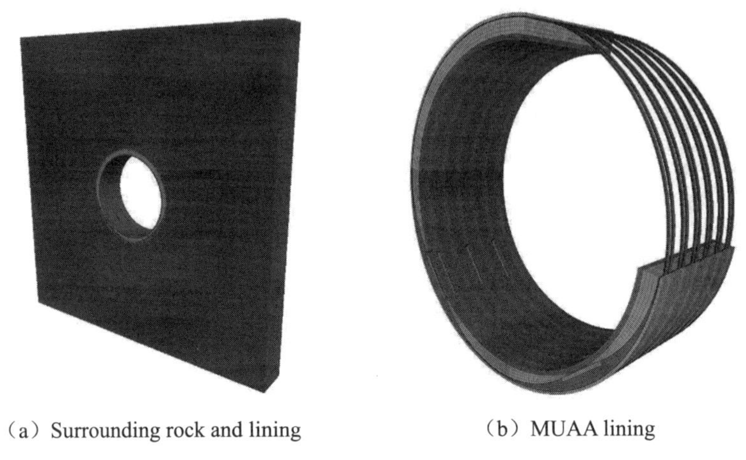

(a) Surrounding rock and lining (b) MUAA lining

Fig. 4.4 Three-dimensional numerical model of the MUAA lining and surrounding rock

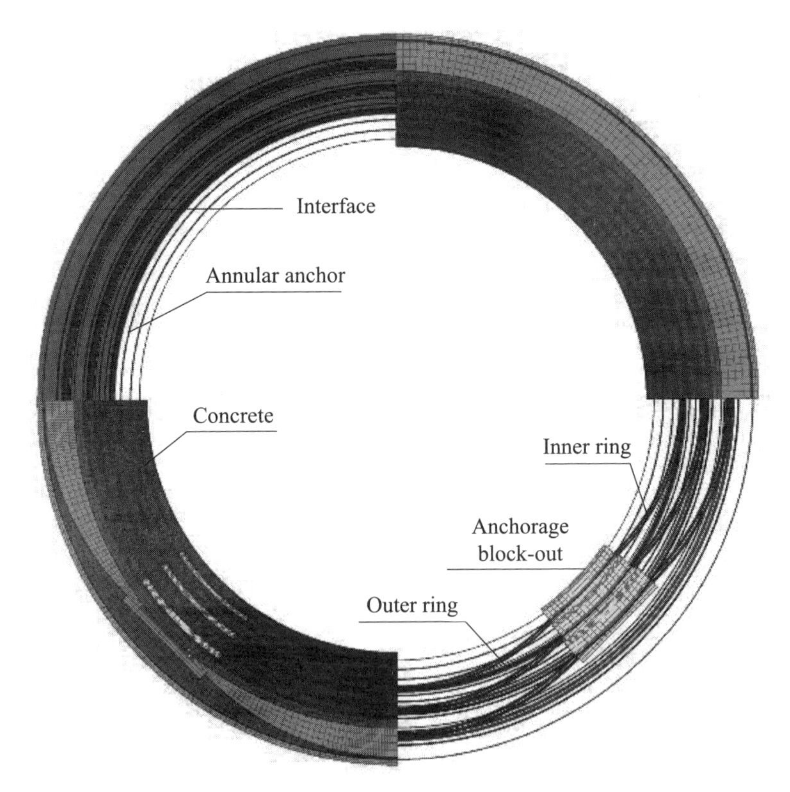

Fig. 4.5 Three-dimensional numerical model of the MUAA lining

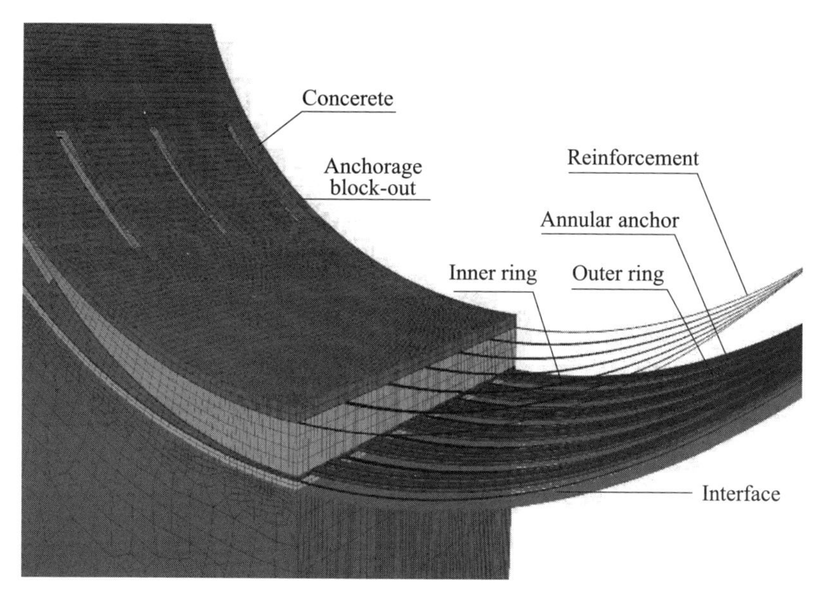

Fig. 4.6 Local numerical model of annular anchor, concrete, and interface

4.4 Mechanical properties by numerical simulation method

4.4.1 Overall prestress

The minimum principal stress (σ_{min}) and maximum principal stress (σ_{max}) of the MUAA lining after tensioning are shown in Fig. 4.7. After the lining is tensioned, the overall prestress effect of the lining is uniform except for the anchorage block-out, and σ_{min} is between 3.0 MPa and 4.5 MPa. The prestress value of the surface parts on the left and right-side walls of the lining is slightly larger, which is between 4.5 MPa and 5.5 MPa.

4.4.2 Prestress distribution of typical section

There are two typical prestress sections along the longitudinal lining, which are the cross section with anchorage block-out (Section A) and the cross section without anchorage block-out (Section B). The σ_{min} and σ_{max} of the MUAA lining after tensioning are shown in Fig. 4.8 and Fig. 4.9. The σ_{min} variation curves of feature points along the wrapping angle are shown in Fig. 4.10.

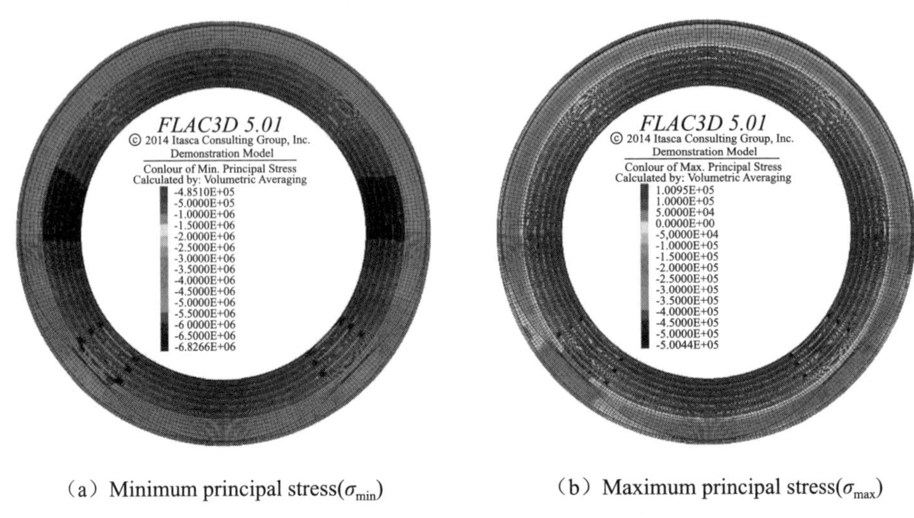

(a) Minimum principal stress(σ_{min}) (b) Maximum principal stress(σ_{max})

Fig. 4.7 Overall prestress of the MUAA lining after tensioning (unit: Pa)

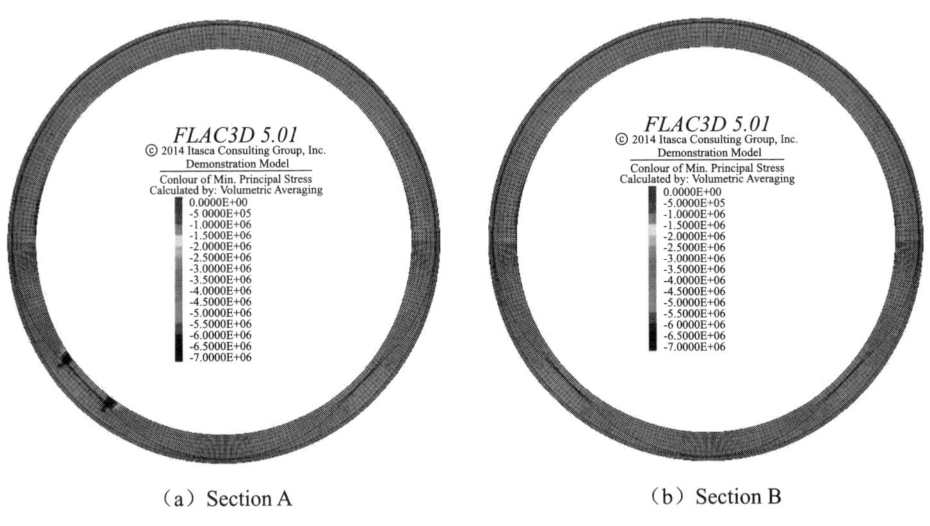

(a) Section A (b) Section B

Fig. 4.8 Minimum principal stress of the MUAA lining after tensioning (unit: Pa)

After the annular anchor is tensioned, the regularity of distribution as well as the magnitude of the σ_{min} of the circumferential prestress of Section A and Section B are basically consistent. The difference between them is less than 5%, both inside and outside of the lining, except near the anchorage block-out. Due to the nonlinear distribution of the friction loss of the annular anchor, the prestress vacancy of the anchorage block-out and the influence of gravity, the distribution of prestressing around 360° is

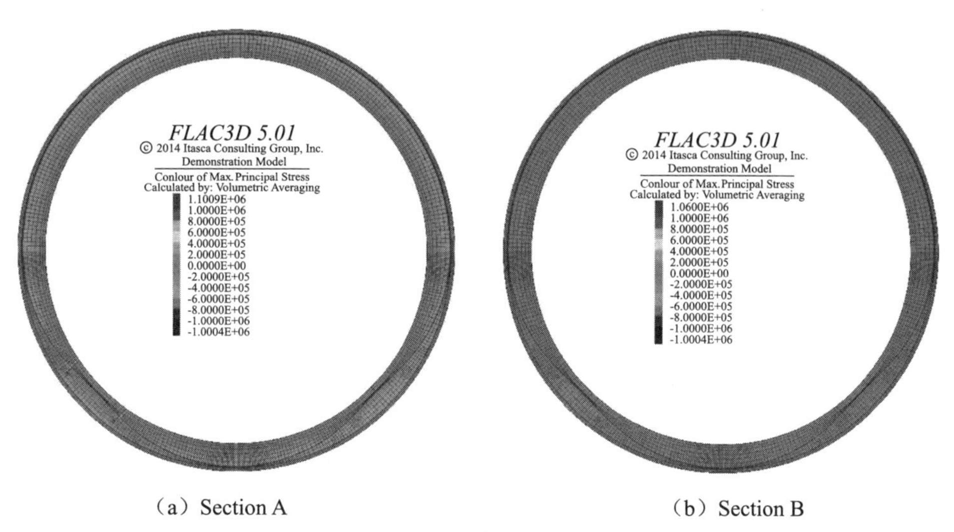

(a) Section A (b) Section B

Fig. 4.9 Maximum principal stress of the MUAA lining after tensioning (unit: Pa)

slightly non-uniform. The maximum prestressing inside of lining is 5.67 MPa, located near 90° and 270°. The maximum prestressing of the outside of lining is 4.44 MPa, located near 135° and 225°.

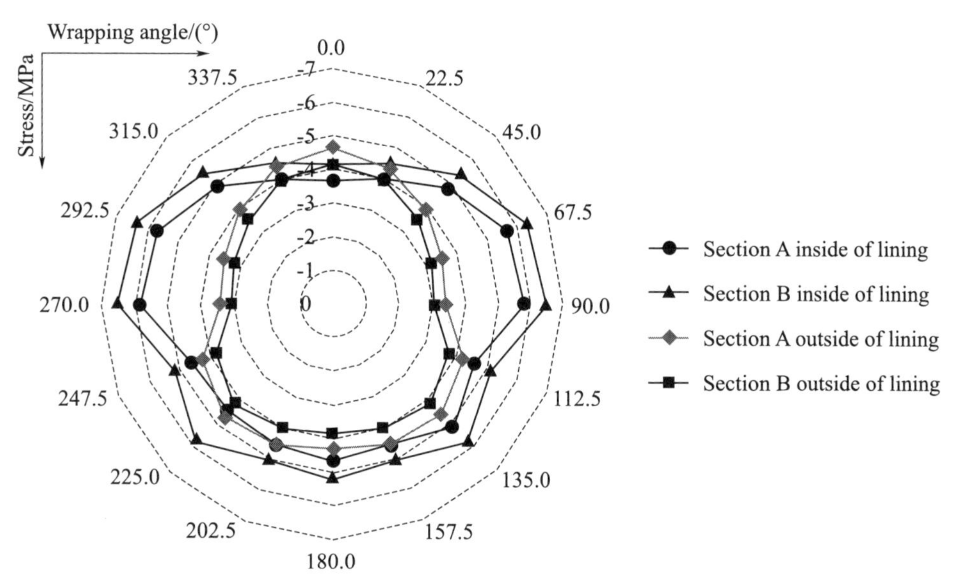

Fig. 4.10 Distribution curves of circumferential prestress inside and outside of lining (unit: MPa)

4.4.3 Lining prestress near anchorage block-out

Due to the existence of anchorage block-out, the local prestress will be lost at 45° below the left and right sides of the lining's inside, and the force near the anchorage block-out is complex. The lining tensile stress area is distributed near the annular surface of the anchorage block-out, and the maximum lining compression area is distributed between adjacent anchorage block-out. The σ_{min} and σ_{max} of the lower-part MUAA lining after tensioning is shown in Fig. 4.11.

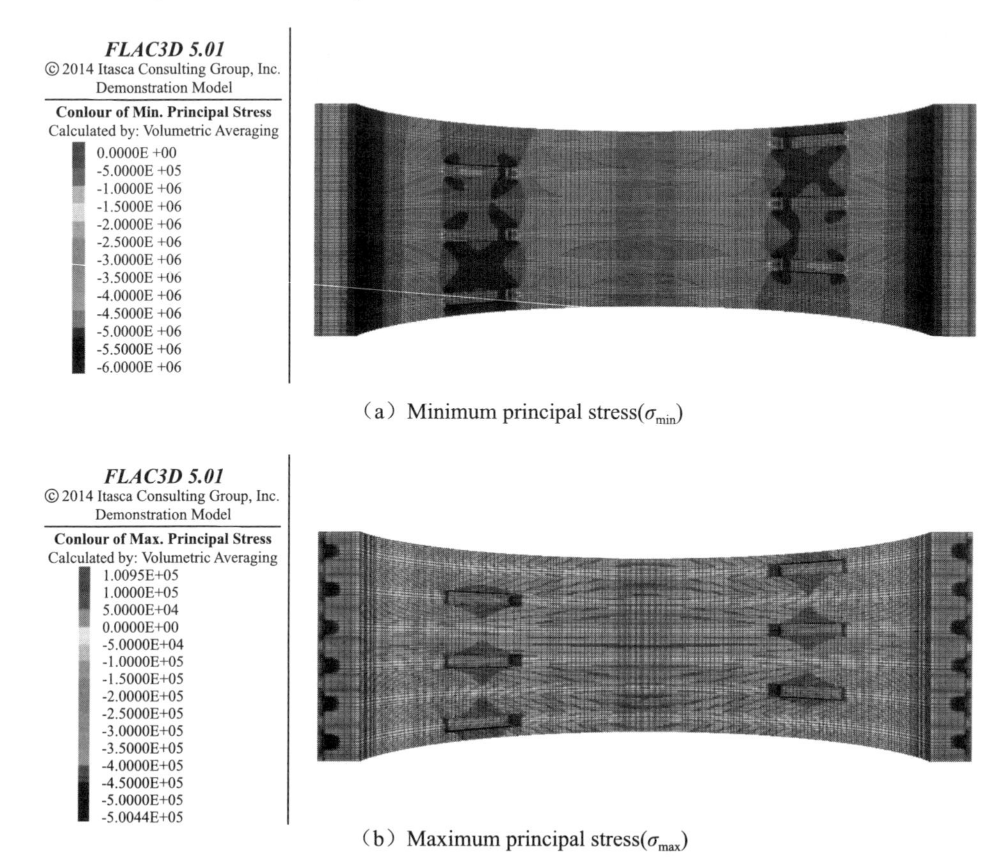

(a) Minimum principal stress(σ_{min})

(b) Maximum principal stress(σ_{max})

Fig. 4.11 The σ_{min} and σ_{max} of the lower-part MUAA lining after tensioning (unit: Pa)

After the annular anchor is tensioned, the maximum circumferential compressive stress (Fig. 4.12) near the anchorage block-out is 7.85 MPa, and there is compressive stress all along the circumferential direction. The maximum longitudinal compressive stress (Fig. 4.13) near the anchorage block-out is 5.72 MPa, and the maximum longitudinal tensile stress is

0.16 MPa. The lining tensile stress area is mainly distributed near the circumferential free face of the anchorage block-out, and the maximum lining compression area is distributed between adjacent anchorage block-outs.

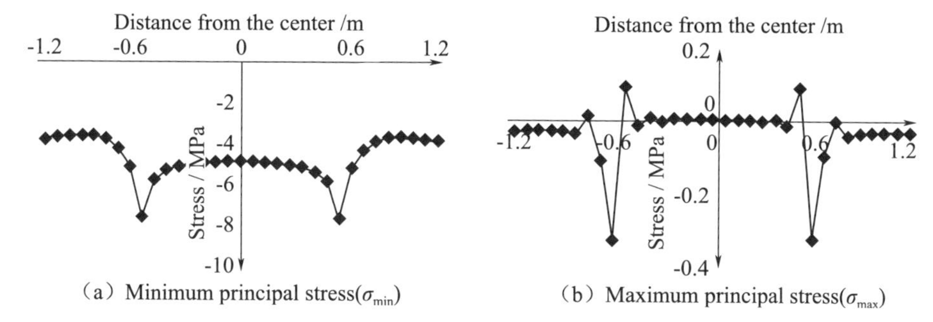

(a) Minimum principal stress(σ_{min})　(b) Maximum principal stress(σ_{max})

Fig. 4.12　Circumferential stress curves near anchorage block-out

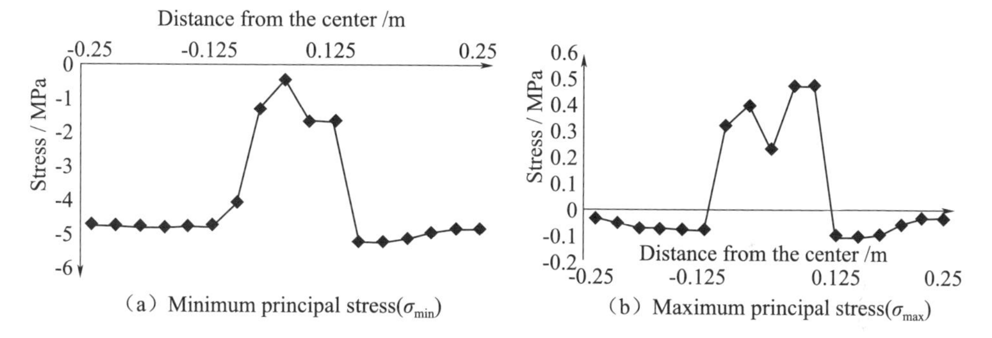

(a) Minimum principal stress(σ_{min})　(b) Maximum principal stress(σ_{max})

Fig. 4.13　Longitudinal stress curves near anchorage block-out

4.4.4　Stress state of reinforcement

After the MUAA lining is tensioned, all the inner and outer non-prestressed reinforcement are in a state of compression. As can be seen from the Fig. 4.14, the reinforcement behind the anchorage block-out is more compressed, with the maximum compressive stress value reaching 33.71 MPa, distributed around 135° and 225° of the wrapping angles.

4.4.5　Deformation of the MUAA lining

Due to the tensioning, the prestress makes the concrete in the middle and upper part shrink inward, and the lining separates from the surrounding rock boundary, also, the opening of the joint gradually increases. After

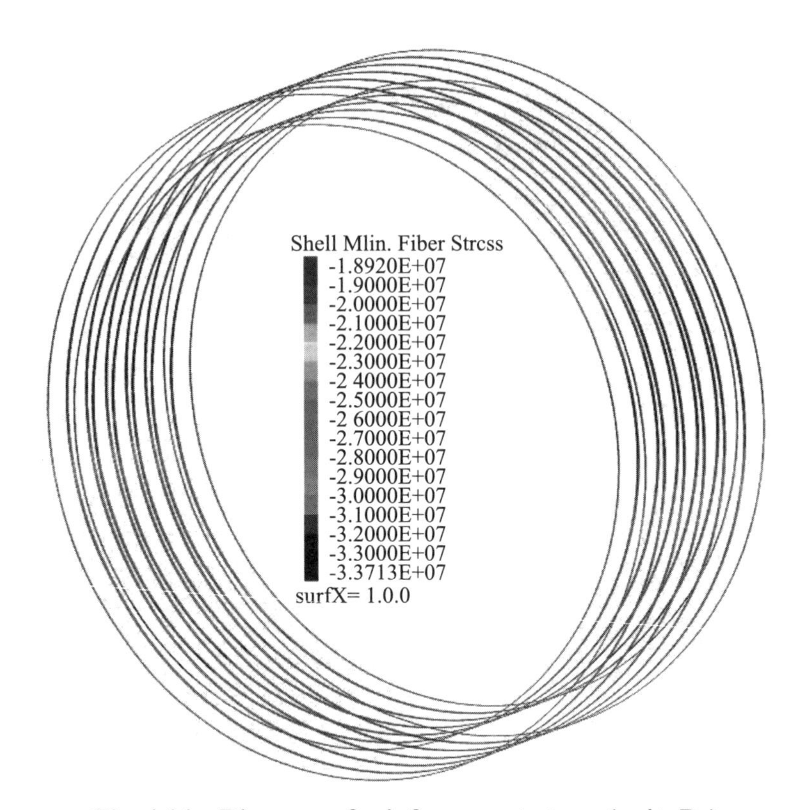

Fig. 4.14 Diagram of reinforcement stress (unit: Pa)

tensioning, the joint opening of interface (Fig. 4.15) between the MUAA lining and surrounding rock reaches a maximum of 1.6 mm. Therefore, engineers can use backfilling or contact grouting for the lining to promote the combined loading capacity, and increase the safety margin of the MUAA lining.

4.4.6 Capacity of the MUAA lining during the usage period

The principal stress ($\sigma_{\min}$ and $\sigma_{\max}$) of the MUAA lining at a water pressure of 0.55 MPa during the usage period are shown in Fig. 4.16. The figures show that at an internal water pressure of 0.6 MPa, the lining is overall compressed, tensile stress only occurs in the anchorage block-out. Since the original numerical model cannot simulate the effect of backfilling micro-expansion concrete into the anchorage block-out, the stress improvement effect after anchorage block-out backfilling cannot be

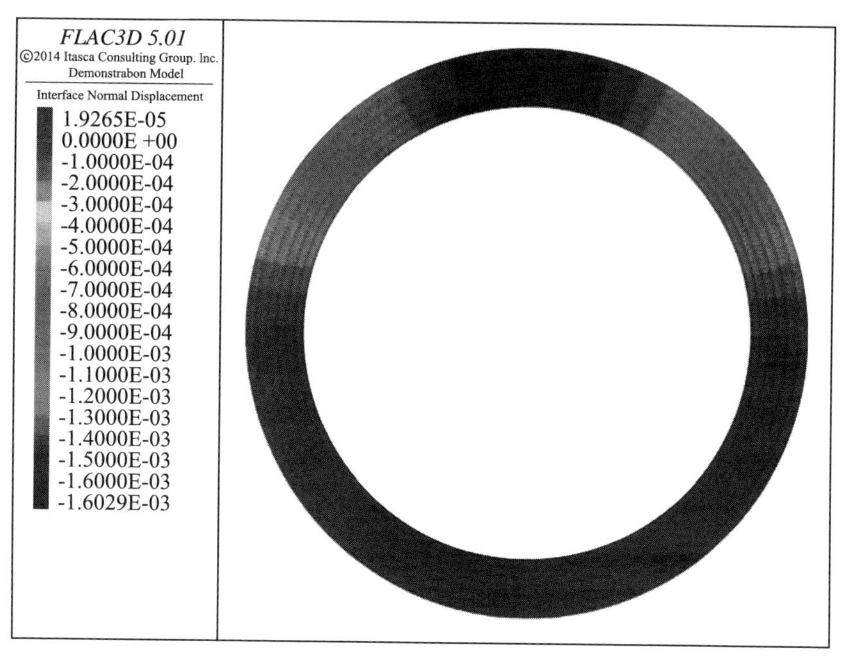

Fig. 4.15 The joint opening of interface between lining and surrounding rock (unit: m)

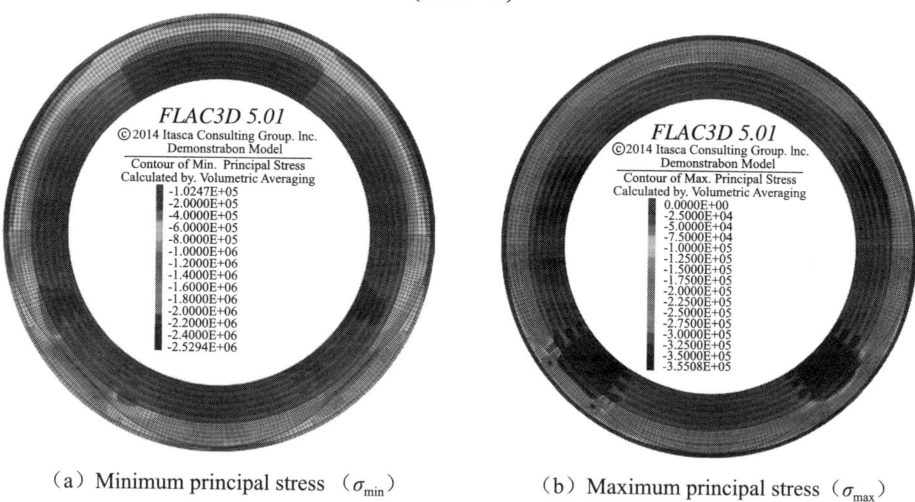

(a) Minimum principal stress (σ_{min}) (b) Maximum principal stress (σ_{max})

Fig. 4.16 Overall prestressing of the MUAA lining during the usage period (unit: Pa)

reflected. The overall stress distribution of the lining is relatively uniform during the usage period, and the circumferential precompression stress is mostly between 0.5 MPa and 2.0 MPa. Except for the anchorage block-out part, the inner surface of the left and right sides of the lining shows the highest precompression stress value. The overall stress distribution of the lining is very similar to that of the stress distribution after tensioning, which is completely consistent with the field test results. In the radial direction, the lining precompression stress at the top and its vicinity shows a gradient decline from the outside to the inside, while the lining precompression stress at other parts gradiently increases.

4.5 Validation of numerical model

The modeling method proposed above is used to calculate the stress and strain state of the MUAA lining, and the reliability of the numerical calculation results is verified by comparing the numerical calculation results with the actual monitoring data.

4.5.1 Lining prestressing

According to the comparison diagrams (Fig. 4.17) of prestress values between the monitoring data and the numerical simulation results, the

- ■ Monitoring values of outside lining by steel strain meter
- ▲ Monitoring values of outside lining by concrete strain meter
- ◆ Numerical calculation results of outside lining
- ● Monitoring valucs of insidc lining by stccl strain mctcr
- ▼ Monitoring values of inside lining by concrete strain meter
- ◁ Numcrical calculation results of inside lining

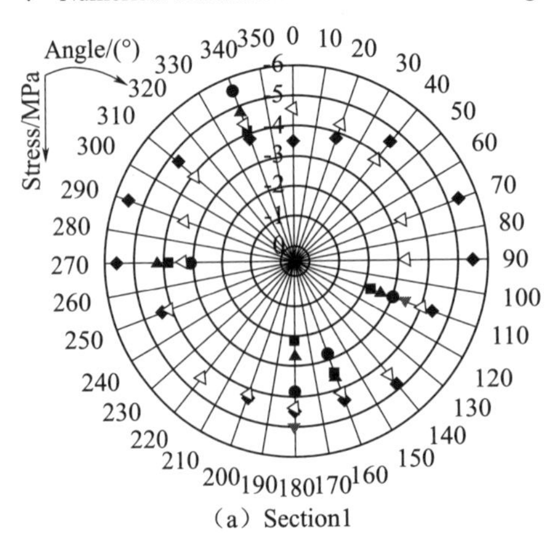

(a) Section1

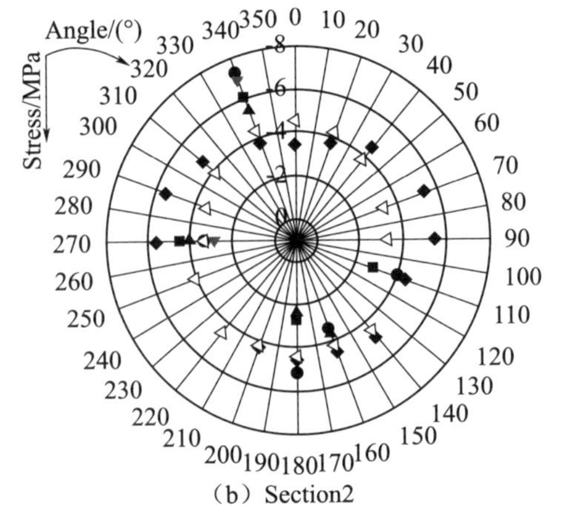

(b) Section2

Fig. 4.17(1) Monitoring values and numerical simulation results

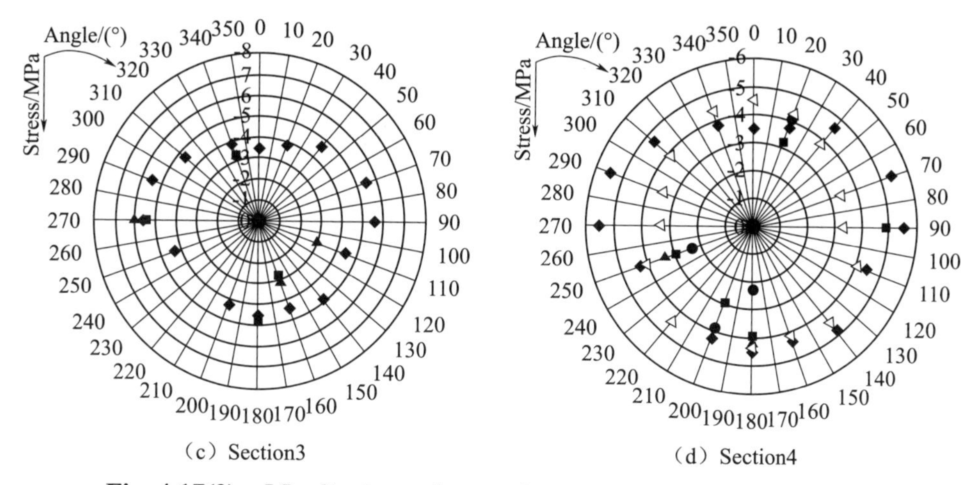

Fig. 4.17(2) Monitoring values and numerical simulation results

numerical calculation results are in good agreement with the measured data in terms of lining prestress distribution regularity. Along the direction of lining thickness, the prestress value decreases from both sides to the top and bottom. Except for the stress concentration in the concrete near the anchorage block-out, the overall prestress distribution is uniform.

4.5.2 Stress state near anchorage block-out

We use strain gauges to monitor the stress of the concrete around the anchorage block-out, and Surfer 13.0 program to draw the stress contour map (Fig. 4.18). The numerical calculation results are basically consistent with the rules of the basic response of the field monitoring.

The anchorage block-out and its surrounding area are weak areas of prestressing effect, and the prestressing values at both ends of the anchorage block-out length direction (tensioning end and anchoring end) are low.

The weak lining areas are mainly distributed near the free face of the circumferential anchorage block-out, and areas with the largest lining pressure are distributed between adjacent anchorage block-outs. The maximum compressive stress value calculated by numerical method is 7.85 MPa, and the field monitoring value is 8.21 MPa. The stress state regularities of anchorage block-out are also basically consistent.

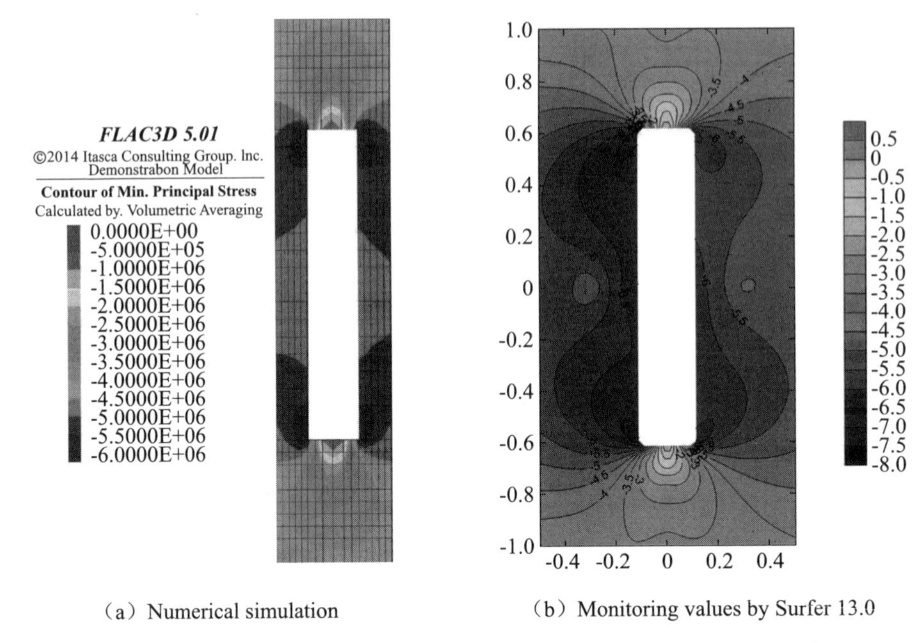

(a) Numerical simulation (b) Monitoring values by Surfer 13.0

Fig. 4.18 Stress contour maps of concrete near anchorage block-out

4.5.3 Potential failure mode

Analysis results of the potential failure characteristics of the lining weak areas indicate that there may be two kinds of gaps near the anchorage block-out (Fig. 4.19):

(1) One kind of gap is perpendicular to the circumferential free face of the anchor block-out and develops deeply into the lining along the radial pressure line of the anchor.

(2) The other gap starts from the corner of the rectangular anchorage block-out and expands to the 45° direction of the circumferential free face of the anchor block-out.

The sketch of the gap's distribution in the MUAA lining after tensioning at Xiaolangdi Project (Fig. 4.20) also reflects the phenomenon of gaps caused by stress concentration. After the anchor is fully tensioned, the concrete gaps and even a small number of fragments break out near the anchor block-out inside the lining.

Therefore, after the prestressed system is tensioned, the anchor block-out should be checked. If there are gaps and other problems, it should be carefully backfilled and repaired to make sure that the lining is uniformly stressed as a whole.

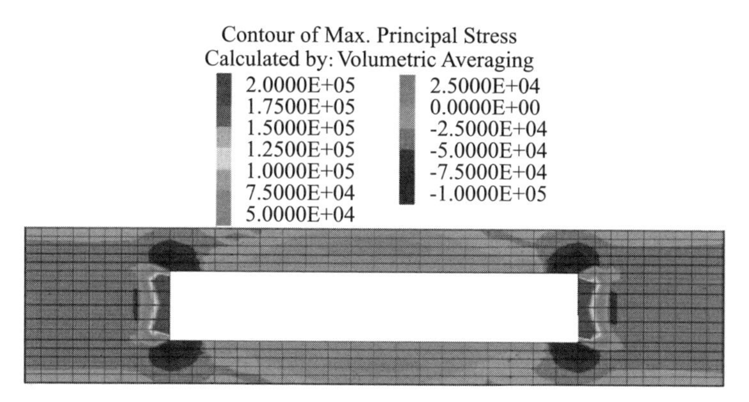

Fig. 4.19 Local maximum principal stress distribution near anchorage block-out (unit: MPa)

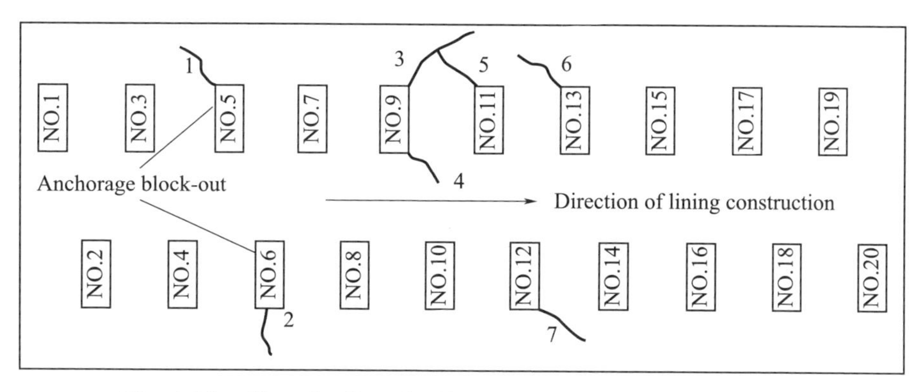

Fig. 4.20 Sketch of gap's distribution in the MUAA lining

4.6 Discussion

The MUAA lining is a new type of structure used in hydraulic pressure tunnel. Firstly, the key points of numerical modeling are analyzed according to the stress characteristics of the MUAA lining. Then, the corresponding modeling methods are proposed: ① the contact relationship is set to simulate the different constraint effects of surrounding rock on the lining,

② the equivalent load and solid model are superposed to simulate the prestress state of the changing anchor, and ③ the nonlinear distribution of prestress loss is simulated by applying gradient load in sections. Finally, based on the finite difference software FLAC3D, an engineer case is carried out to verify the correctness of the modeling method presented. The results show that this new modeling method has clear principle, rapid-prototyping and reliable results, which is of guiding significance to structure design.

References

ANDERSON P, HANSSON M, THELANDERSSON S, 2008. Reliability-based evaluation of the prestress level in concrete containments with unbonded tendons[J]. *Structural Safety*, 30(1): 78-89.

CAO R L, WANG Y J, ZHAO Y F, et al, 2016. Study on numerical modeling method of the prestressed tunnel lining with unbonded curve anchored tendons[J]. *Journal of China Institute of Water Resources and Hydropower Research*, 14(6): 471-477.

DAVID B G, JOS M, DEAN J D, et al, 2016. Prestress Loss Database for Pretensioned Concrete Members[J]. *ACI Structural Journal*, 113(2): 313-324.

FAHIMIFAR A, SOROUSH H, 2005. A theoretical approach for analysis of the interaction between grouted rock bolts and rock masses[J]. *Tunneling and Underground Space Technology*, 20: 333-343.

GHAZAVI M, RAVANSHENAS P, LAVASAN A A, 2014. Analytical and numerical solution for interaction between batter pile group[J]. *KSCE Journal of Civil Engineering*, 7: 1-13.

HAJALI M, ALAVINASAB A, SHDID C A, 2016. Structural performance of buried prestressed concrete cylinder pipes with harnessed joints interaction using numerical modeling[J]. *Tunnelling & Underground Space Technology*, 51(1): 11-19.

KANG J F, LIANG Y H, ZHANG Q C, 2006. 3-D finite element analysis of post-prestressed concrete lining[J]. *Journal of Tianjin University Science and Technology*, 39(8): 968-972.

KANG J F, SUI C E, WANG X Z, 2014. Structure optimization of the prestressed tunnel lining with unbonded circular anchored tendons[J]. *Journal of Hydraulic Engineering*, 45(1): 103-108.

TAHMASEBINIA F, RANZI G, ZONA A, 2012. Beam tests of composite steel-concrete members: a three-dimensional finite element model[J]. *International Journal of Steel Structures*, 12(1): 37-45.

WANG Y J, CAO R L, PI J, et al, 2020. Mechanical properties and analytic solutions of prestressed linings with unbonded annular anchors under internal water loading[J]. *Tunnelling and underground space technology*, 97(3): 1-10.

WANG Q, ZHU L, LI Z, et al, 2013. Study on the constitutive model of steel for explicit dynamic beam elements of Abaqus[J]. *China Civil Engineering Journal*, 46: 100-105.

XIANG YAN, 2014. Influence of temperature stress on internal force and deformation of retaining structures for deep excavations[J]. *Chinese Journal of Geotechnical Engineering*, 36(zk2): 64-69.

Chapter 5
Influence of various factors on mechanical characteristics of the MUAA lining

5.1 Introduction

In order to realize the tensioning of annular anchors, the lining needs to set anchorage block-out in advance to fix the annular anchors before concrete pouring (Bian et al, 2009; Koryt'ko et al, 2005). The anchorage block-out area is the most complex position of the whole lining. Due to the existence of the anchorage block-out, the stress on the lower half of the lining is uneven, and the preloading stress on the lining in this area varies greatly. The shape of the anchorage block-out is usually rectangular, and the stress on the corners is uneven, which is easy to produce corner cracks (Behfarnia et al, 2014; Lee et al, 2007; Yan & Zhu, 2008). After the annular anchors are tensioned, the most unfavorable stress point of the whole lining appears in the anchorage block-out area, and the lining may be subjected to radial tensile stress near the free surface (Jin, 2006). The lining between two adjacent rows of anchorage block-outs may be subjected to great circumferential compressive stress.

The prestressed anchors embedded in the MUAA lining are encircled by the one-layer and two-hoop ways to reduce the lining thickness. A thinner lining can both decrease the construction cost and increase the cross-sectional area of water conveyance to promote the operational efficiency of a tunnel (Cao et al, 2019; Lin & Shen, 1999). Because annular anchors are tied to the inner side of non-prestressed steel bar, the quality of the cast-in-situ concrete will be affected if the lining thickness is rather reduced (Jin, 2005; Kang et al, 2014). Previously, studies demonstrated that the circumferential compressive stress will decrease as the lining thickness increases, which indicates that the thinner the lining, the more obvious the prestressing effect (Simanjuntak et al, 2016). However, with the reduction of lining thickness, the circumferential tensile stress will correspondingly increase because of the internal water pressure. Moreover, the overbreak of tunnel sides generally occurred to meet the traffic demand and the passage of the second-lined steel bar trolley, which will change the lining shape from a circle to an ellipse.

The Drilling-blasting method is one of the main methods of tunnel excavation (Minh et al, 2021; Song et al, 2014). It will inevitably lead to local over or under excavation of surrounding rock. The cross-section shape of tunnel will be non-circular, and the common shape is horseshoe (Yan et al, 2015). The stress of the MUAA lining is complex, and the change of tunnel section shape will further increase the complexity of internal stress state.

Therefore, the influence of various factors on the mechanical characteristics of the MUAA lining are studied by 3D numerical simulation method, including the position of anchorage block-out, lining thickness and cross-section shape.

5.2 Influence of anchorage block-out on mechanical characteristics

5.2.1 Numerical model

The position of anchorage block-out is very important to the prestressing force and safety for the entire MUAA lining. FLAC3D is used for 3D modeling to analyze the stress of key parts when anchorage block-outs are arranged in single row and double rows (Fig. 5.1).

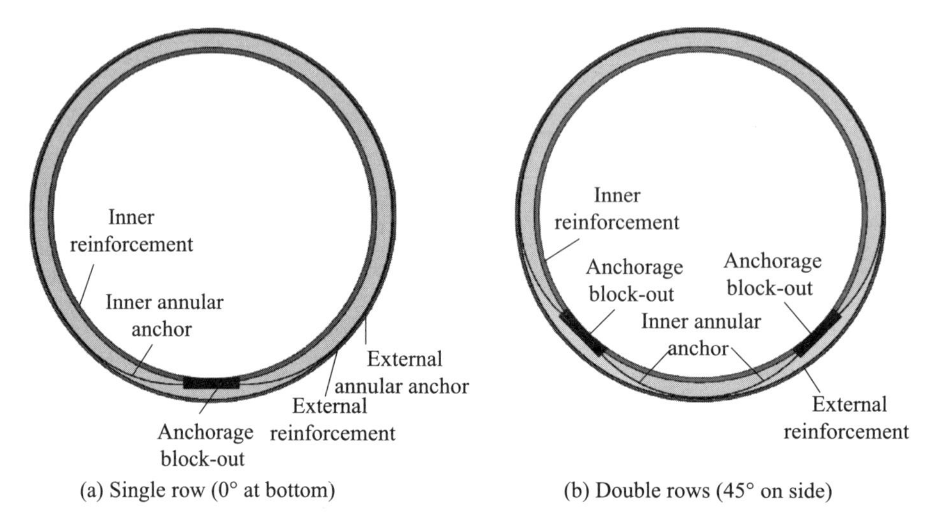

Fig. 5.1 Anchorage block-outs at the bottom and side of the MUAA lining

5.2.2 Influence of the position of anchorage block-out on mechanical characteristics

After the annular anchors are tensioned, the stress of the lining with different position of anchorage block-out is shown in Fig. 5.2 and Fig. 5.3. The prestress of the upper half ring of the lining is evenly distributed, while the local prestress of the lower half ring of the lining will be missing and unevenly distributed due to the influence of the anchorage block-out.

There is obvious stress concentration in the concrete near the anchorage block-out along the axial and circumferential directions, and the maximum compressive stress is 8.5-9.0 MPa, the maximum tensile stress is 0.5-0.7 MPa. When the single row anchorage block-out is arranged, the overall prestress of the lining will be more uniform at the lower half ring of the lining.

It can be seen from the stress curves of the MUAA lining near anchorage block-out in Fig. 5.4 that after the annular anchors are tensioned, the most unfavorable stress position of the lining is near the anchorage

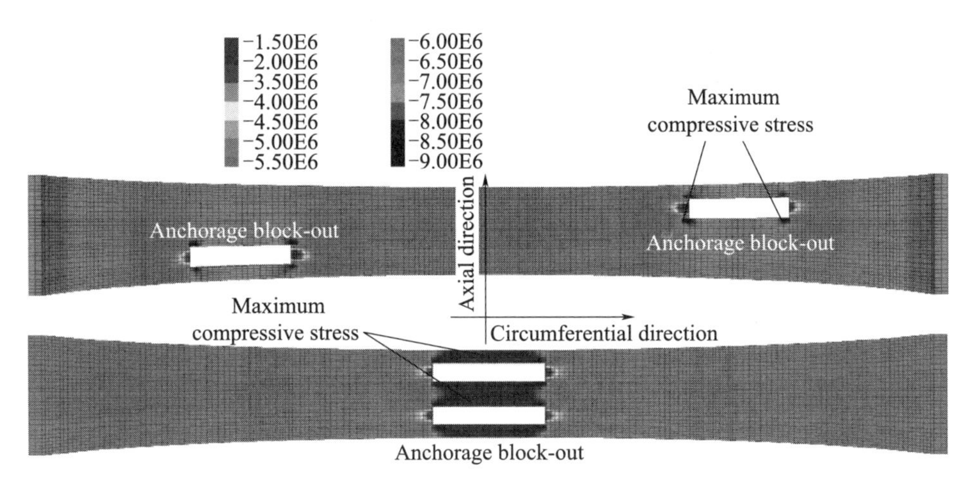

Fig. 5.2 Minimum principal stress of the MUAA lining after tensioning (unit: Pa)

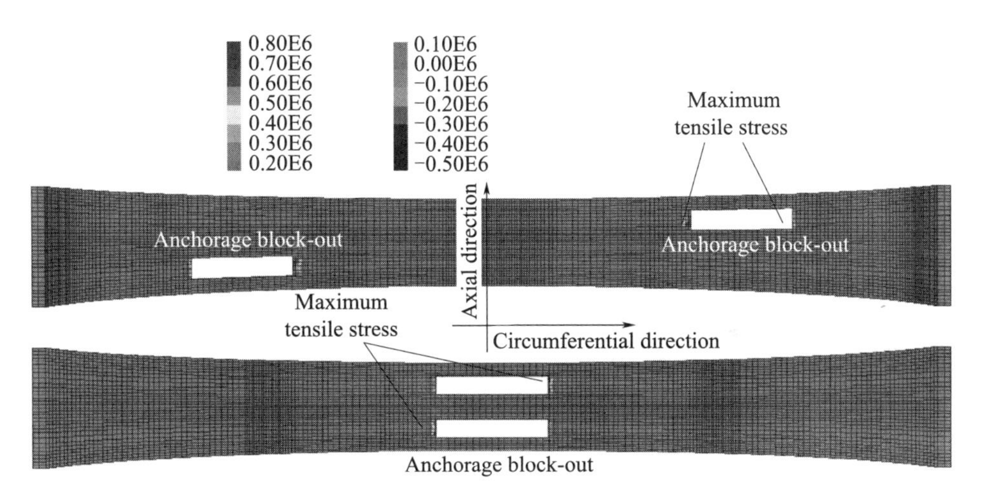

Fig. 5.3 Maximum principal stress of the MUAA lining after tensioning (unit: Pa)

block-out. The tensile stress area of the lining is mainly distributed near the free face in the circumferential direction of the anchorage block-out, and the area with the greatest pressure on the lining is distributed between the adjacent anchorage block-outs. The reinforcement of this area should be strengthened in the design. Compared with the arrangement of double row anchorage block-outs, the compressive stress in the arrangement of single row is more concentrated, and the maximum difference is 2.5 MPa, mainly because the concrete stress state in this part is the result of the superposition of the effects of two adjacent anchorage block-outs.

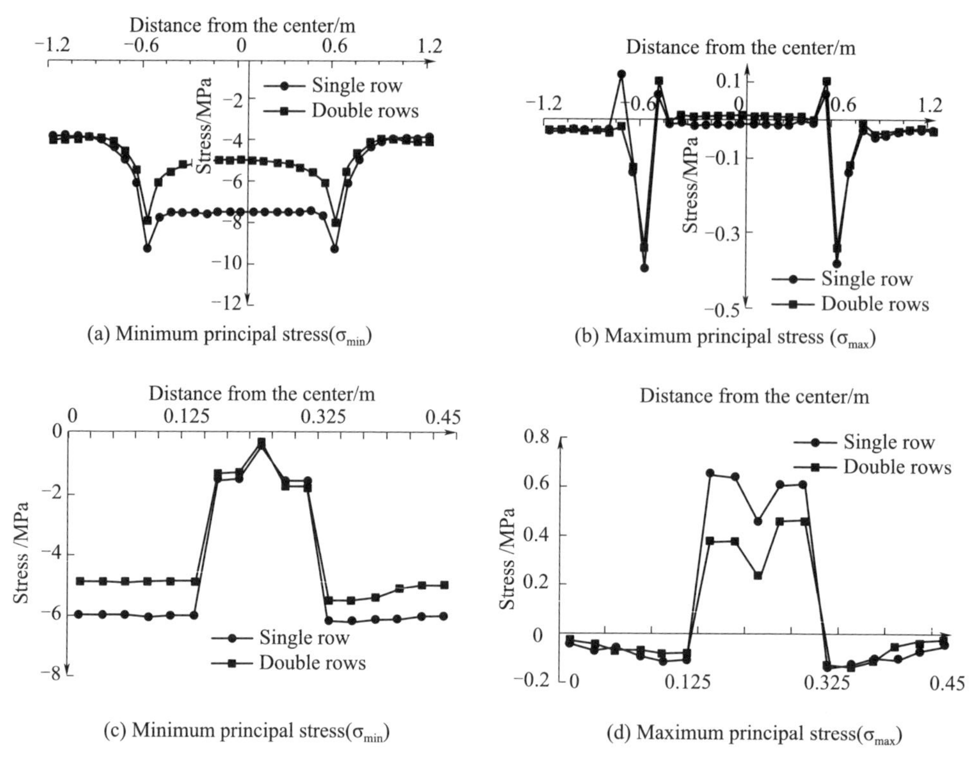

(a) Minimum principal stress(σ_{min})

(b) Maximum principal stress (σ_{max})

(c) Minimum principal stress(σ_{min})

(d) Maximum principal stress(σ_{max})

Fig. 5.4 Stress curves of the MUAA lining near anchorage block-out (unit: Pa)

The stress state of the MUAA lining under 55 m (pressure head during operation), 65 m, 75 m and 100 m overpressure head is calculated respectively. The calculation results are shown in Table 5.1. The unfavorable stress position of the two types is the anchorage block-out area. When bearing overload water pressure, the lining concrete of single row anchorage block-out is more likely to be damaged. For example, when the overload water pressure is 0.75 MPa, the lining runs through the tensile stress zone along the longitudinal direction of the anchorage block-out. However, in case of double row anchorage block-outs, the overall compressive stress of the lining is about 0-0.33 MPa, which is still in a safe state. Therefore, the internal water pressure resistance of the double row anchorage block-outs scheme is better than that of the single row.

Table 5.1 The stress state of the MUAA lining near the anchorage block-outs

Pressure head /m	Maximum compressive stress /MPa		Maximum tensile stress /MPa	
	Single row	Double rows	Single row	Double rows
55	0.67-1.73	0.75-1.95	2.3	2.2
65	0.45-0.78	0.65-0.89	2.7	2.7
75	None	0-0.33	3.2	2.9
100	None	None	4.3	3.2

5.3 Influence of lining thickness on mechanical characteristics

5.3.1 Numerical model

The numerical models of the MUAA lining with variable thickness are shown in Fig. 5.5. This study focuses on investigating the stress variation of the lining with an average thickness of 45, 50, 55, and 60 cm during construction and operation period through a 3D numerical modeling method. Furthermore, we discussed the stress spatial distribution characteristics and the weak loading area of the lining with elliptical casting section. Moreover, we studied the influence of lining thickness on its mechanical properties.

The concrete grade of the MUAA lining section is C40. The internal diameter is 6.8 m. The prestressed annular anchors are wrapped by a single-layer double-circle method with a spacing of 0.5 m. The standard value of the tensile strength is 1860 MPa, the friction coefficient of the steel strand is 0.032, and the swing coefficient is 0.0007. The anchorage block-out is set at the 45° left and right lower sides of the lining, and its length, width, and depth are 1.20, 0.20, and 0.20 m, respectively. Because of the complexity of the model and many elements, if the seepage force applies the external water load, the model will slowly converge during the numerical calculation process. Therefore, the external water pressure is applied by the equivalent force on the surface.

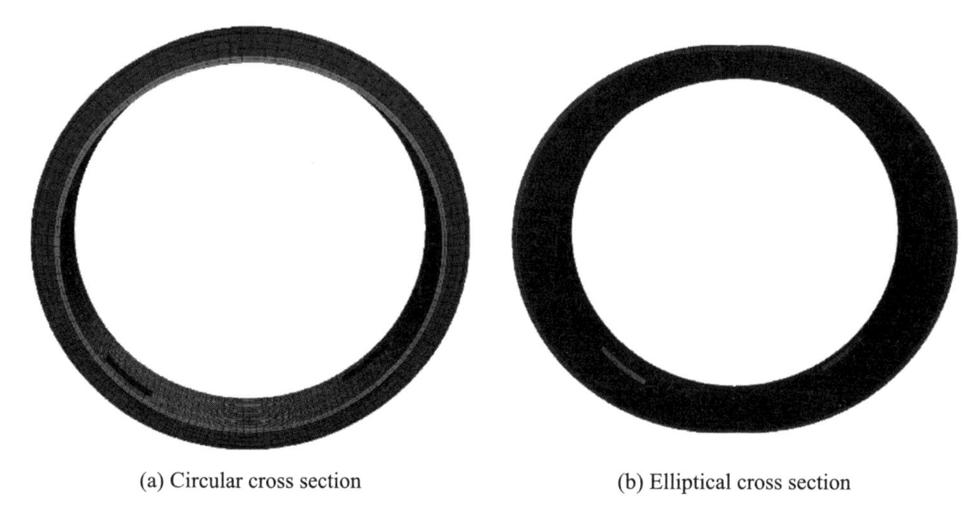

(a) Circular cross section (b) Elliptical cross section

Fig. 5.5 The numerical model of the MUAA lining with variable thickness

5.3.2 Influence of the average thickness on mechanical properties

With a 50 cm thickness, the overall prestressing effect of the lining is uniform. The primary circumferential prestress is between 3.0 MPa and 4.5 MPa. The prestress on the surface of the left and right-side walls of the lining is slightly larger between 4.5 MPa and 5.5 MPa. As shown in Fig. 5.6, the prestress gradually increases as the thickness of the lining decreases. If the thickness decreases by 5 cm, the overall prestress increases by 0.3-0.5 MPa. The prestress distribution pattern remains basically consistent with the change of the lining thickness. Overall prestressed distribution is that the maximum prestress is at the sidewall and that of the vault is slightly lower.

Because of the anchorage block-out, there is local prestress loss at the 45° left and right lower sides of the lining. Moreover, the load status close to the tension block-out is complicated. The tensile stress area of the lining distributes close to the circumferential free surface of the anchorage block-out, whereas the maximum compressive stress area distributes at the section between the adjacent anchorage block-outs. With a 50 cm lining thickness, the maximum compressive stress near the anchorage block-out is 7.16 MPa and the maximum tensile stress is 1.13 MPa. With a 55 cm lining thickness, the maximum compressive stress near the anchorage block-out is 6.45 MPa and the maximum tensile stress is 1.08 MPa (Fig. 5.7). The lining thickness has little effect on the stress of the lining concrete close to the anchorage block-out.

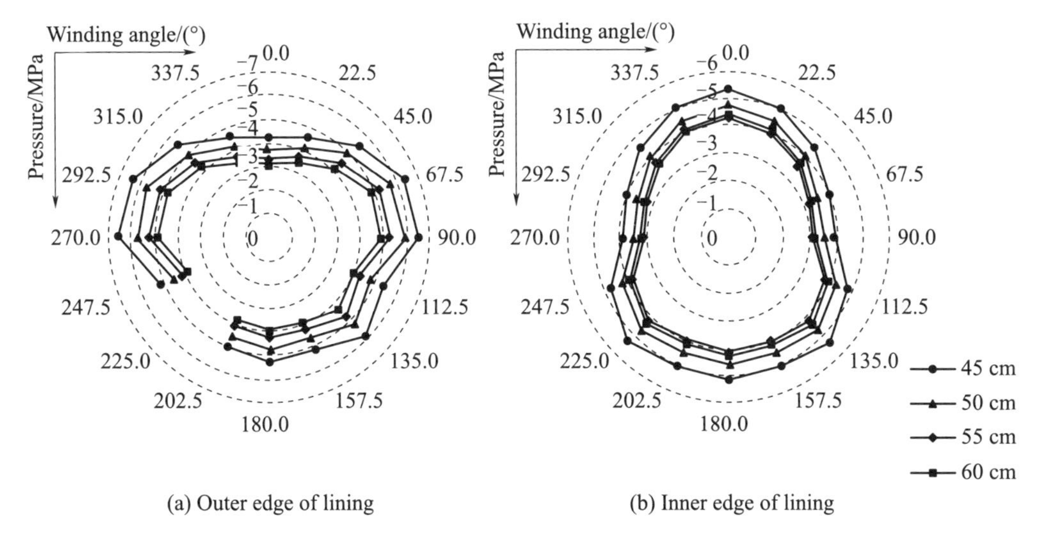

(a) Outer edge of lining

(b) Inner edge of lining

Fig. 5.6 Prestress circumferential distribution of the lining (unit: MPa)

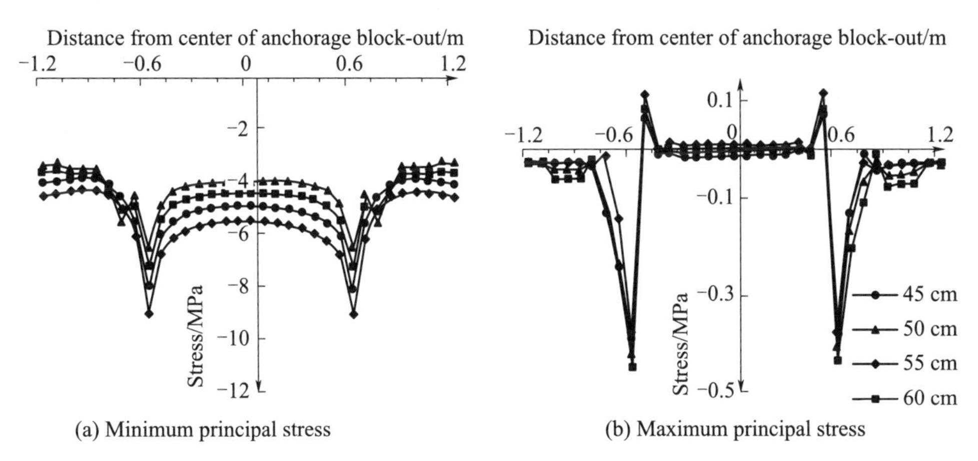

(a) Minimum principal stress

(b) Maximum principal stress

Fig. 5.7 Prestress circumferential distribution of the anchorage block-out (unit: MPa)

5.3.3 Influence of local thickness changes on mechanical properties

To meet the traffic demand during the construction and the passage of the second-lined steel bar trolley, the over-excavation is larger in the horizontal direction of the hance of local tunnel segments. Steel bars and annular anchors are installed as an ellipse. The thickness of the protective layer of reinforcing steel on the inner side of the lining increases from 5 cm to 34 cm. It is significant to clarify the influence of such variation on the mechanical properties of the lining.

(1) Comparison of minimum principal stress of the lining after tensioning. Stress within the MUAA lining will redistribute with the action of tensioning. From the contour of compressive stress in Fig. 5.8 and Fig. 5.9, the compressive stress distribution characteristics of the structure are distinct with the local thickness changing.

As the lining thickens, the overall prestress of the lining reduces. The average prestress is 5.5 MPa and 4.3 MPa for the circular lining and elliptical lining, respectively. The difference between the two values is 1.2 MPa, i.e., 21.8%. At the largest thickness section of the lining on both sides, the prestress is 5.3 MPa for the circular lining and 3.9 MPa for the elliptical lining. The difference between the two values is 26.4%.

The prestress at the crown and invert of the circular lining distributes uniformly. Because of the different thickness, the elliptical lining leads to the axial pressure and bending moments on the structure. The stress concentration exists at the crown and invert of the lining. The concrete at the invert of the lining is dense, and the structure is relatively safe. The crown of lining must be dense through the backfill grouting or consolidation grouting; otherwise, the crown section may develop into a potential damage area.

The stress near the anchorage block-out is often uneven. As the lining thickening, the uneven degree of the stress near the anchorage block-out of the elliptical lining decreases. Therefore, changes in the thickness of the lining will not harm the safety of the anchorage block-out and its nearby concrete.

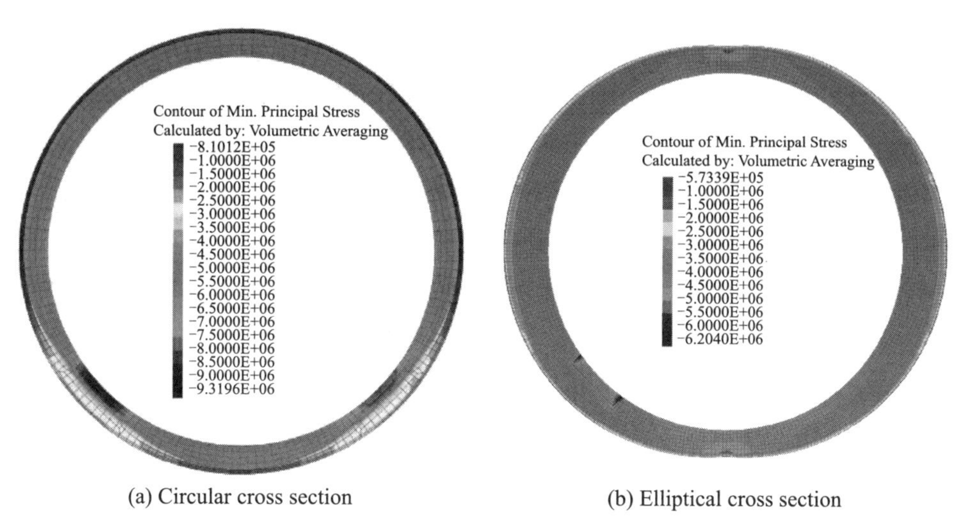

(a) Circular cross section (b) Elliptical cross section

Fig. 5.8 Minimum principal stress of the lining after tensioning (unit: Pa)

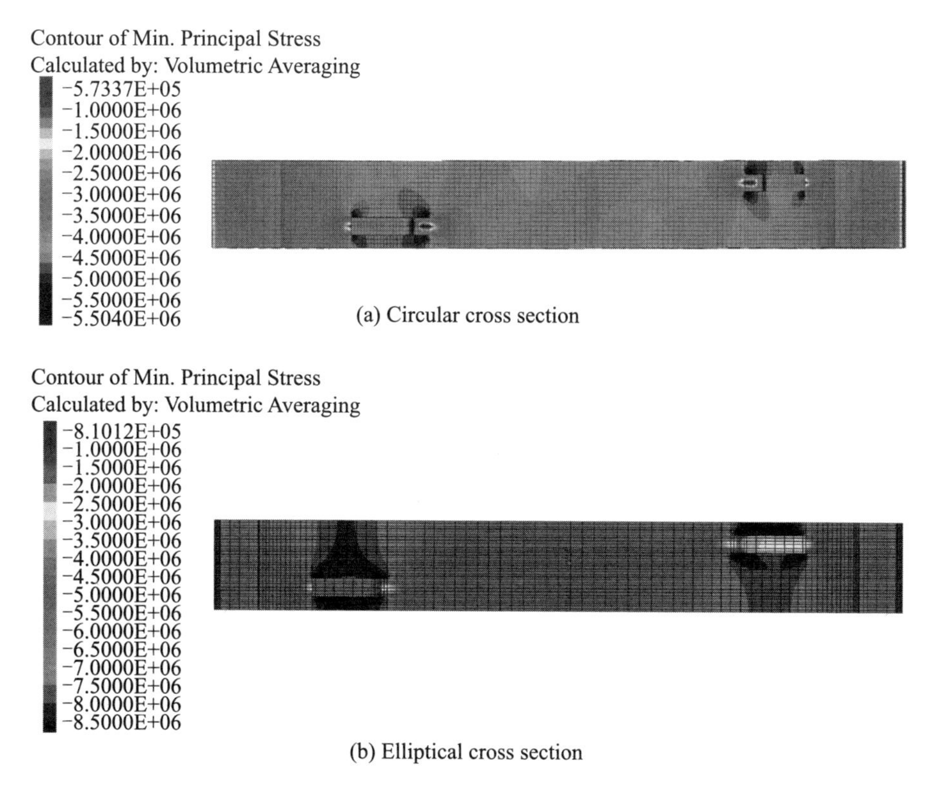

(a) Circular cross section

(b) Elliptical cross section

Fig. 5.9 Minimum principal stress of the lining near anchorage block-out (unit: Pa)

(2) Comparison of maximum principal stress of the lining after

tensioning. After annular anchors being tensioned, the evaluation of lining quality is considerably dependent on its prestressing status. Moreover, tension strength value and distribution characteristics of the concrete of the lining are very important. The tensile stress of the concrete must be less than the allowable value. Fig. 5.10 shows the maximum principal stress distribution. The lining is under compression integrally. In addition to the section of annular anchors, there is rarely a tensile stress zone in the circular lining and elliptical lining. At the larger thickness section of the elliptical lining, there is an annulus [blue in Fig. 5.10(b)] in the middle, all of which is compression stress. There is no tensile stress area and the structure is safe. However, the crown and invert of the elliptical lining exist slight tensile stress (0–0.4 MPa), which is consistent with the above compressive stress analysis results. This indicates that the crown and invert section are still adverse areas.

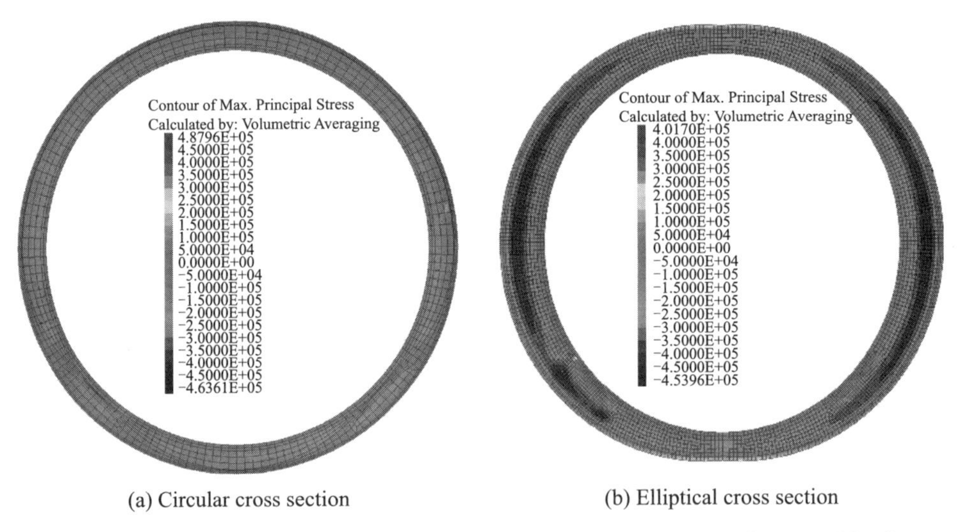

(a) Circular cross section (b) Elliptical cross section

Fig. 5.10 Maximum principal stress of the lining near the anchorage block-out (unit: Pa)

(3) Safety of the lining after applying internal water pressure. After applying internal water pressure, the compressive stress of the lining

gradually decreases, whereas the tensile stress correspondingly increases. The stress state of the prestressed lining has significantly changed. The major distinction between the circular lining and elliptical lining is that, under the internal water pressure, the overall prestress margin of the circular lining is 1.30 MPa, which indicates that the circular lining is safe. When the elliptical lining is only 0.35 MPa, it indicates that the elliptical lining is less safe than the circular lining. After applying internal water pressure, the overall maximum tensile stress is 0.2 MPa for the circular lining and 0.45 MPa for the elliptical one. There is no large tensile stress zone for both, and no obvious damage is reported in the elliptical lining. The non-central symmetry of the structure leads to the bending moment in the elliptical lining and a tensile stress of 1.0 MPa on the inner side. Cracks are more prone to occur on the surface of the elliptical lining, which has a potentially adverse effect on the durability of the structure.

5.4 Influence of cross-section shape on mechanical characteristics

5.4.1 Numerical model

If the spatial geometry of horseshoe lining is symmetrical from left to right and asymmetric from top to bottom, the mechanical properties of horseshoe lining are obviously different from that of circular lining. Therefore, it is necessary to focus on the mechanical properties such as the overall effect of prestress and the concentration of local prestress. The comparative analysis of circular and horseshoe lining is carried out to clarify the influence of tunnel type on the mechanical characteristics of the MUAA lining. The numerical calculation models of the MUAA lining are shown in Fig. 5.11.

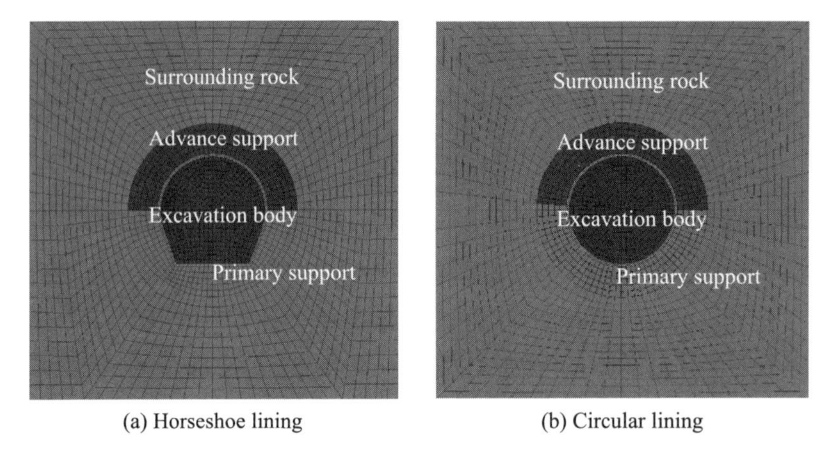

(a) Horseshoe lining (b) Circular lining

Fig. 5.11 Numerical calculation models of the MUAA lining

5.4.2 Influence of cross-section shape on mechanical characteristics

(1) Mechanical characteristics of lining during anchors tensioning. The stress comparison between horseshoe lining and circular lining after anchors tensioning is shown in Fig. 5.12 and Fig. 5.13. The compressive stress of the upper part of the lining is close and evenly distributed, mainly between 4.0-6.0 MPa. For the lower half lining, the overall prestress value is about 10% less (about 0.5 MPa) due to the large area shared by the horseshoe

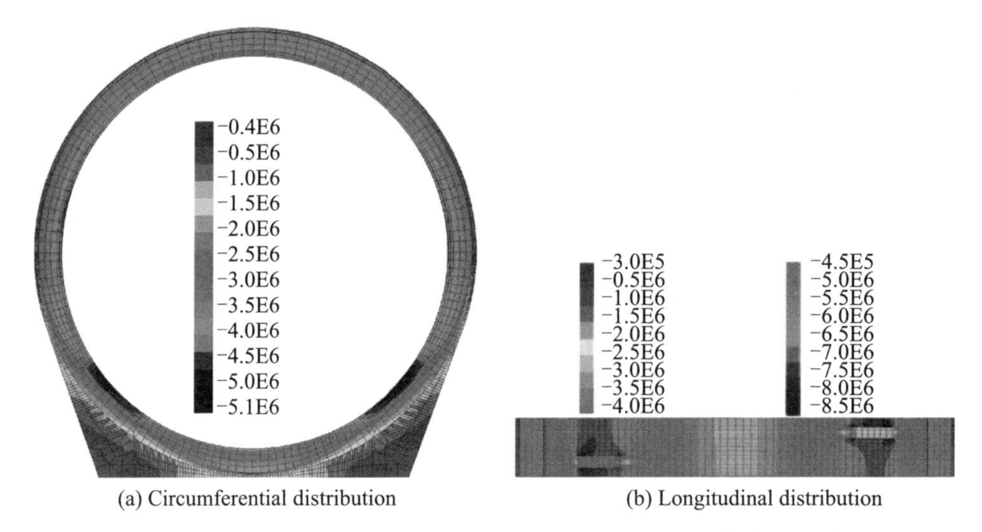

(a) Circumferential distribution (b) Longitudinal distribution

Fig. 5.12 Minimum principal stress of the horseshoe lining (unit: Pa)

lining. In the area with large compressive stress between anchorage block-outs, the maximum compressive stress of horseshoe lining is 8.61 MPa, and the circular lining is 11.33 MPa at the same part. The overall stress of horseshoe lining is more uniform.

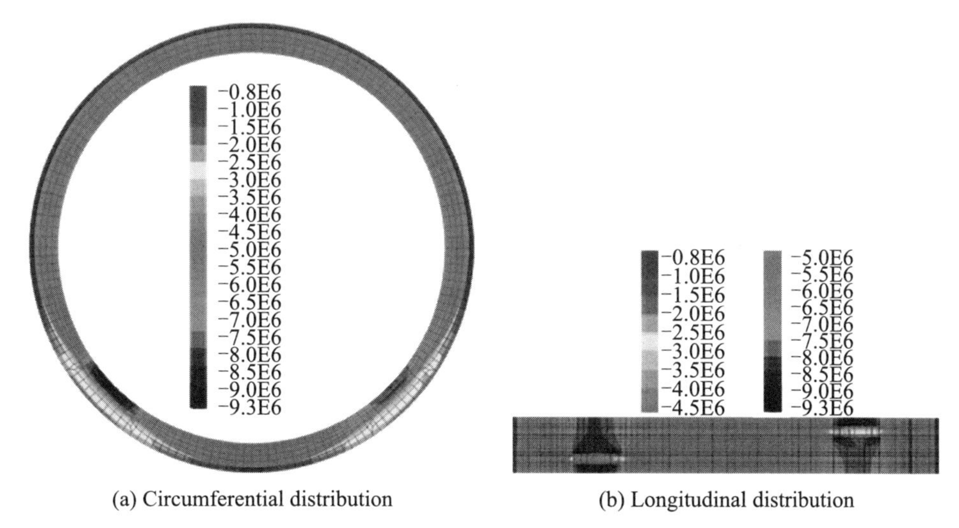

(a) Circumferential distribution (b) Longitudinal distribution

Fig. 5.13 Minimum principal stress of the circular lining (unit: Pa)

(2) Mechanical characteristics of lining during internal water loading. The distribution law of minimum principal stress of lining after internal water pressure is shown in Fig. 5.14 and Fig. 5.15. Compared with the circular lining, the stress of the upper half of the horseshoe lining is not different, and the prestress is about 1.5 MPa. However, the stress state of the lower half of the horseshoe lining is slightly better. On the one hand, between adjacent anchorage block-outs, the stress of horseshoe lining is more uniform and the regional stress difference is smaller. On the other hand, the compressive stress behind the anchorage block-out of the horseshoe lining is relatively large, which can effectively curb the penetration of the tensile stress zone and is conducive to the safety of the lining structure. In general, the structural stress of horseshoe lining is better than that of circular lining.

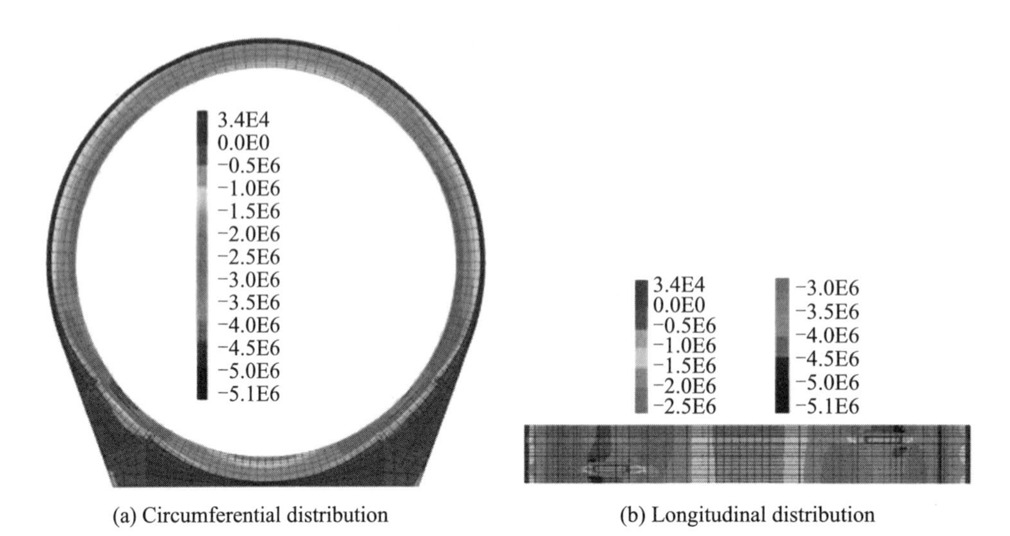

(a) Circumferential distribution (b) Longitudinal distribution

Fig. 5.14 Maximum principal stress of the horseshoe lining (unit: Pa)

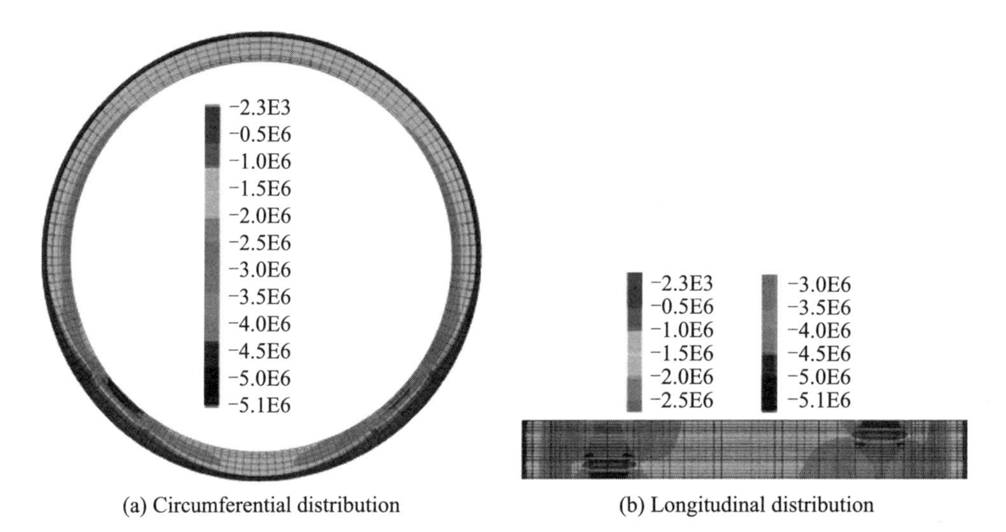

(a) Circumferential distribution (b) Longitudinal distribution

Fig. 5.15 Maximum principal stress of the circular lining (unit: Pa)

(3) Comparative analysis of stress of horseshoe lining and circular lining. See Table 5.2 for the comparison results of mechanical properties of the MUAA linings with different cross-section shape. Although the prestressed effect of horseshoe lining is not different from that of circular lining, the stress distribution at the anchorage block-out is better after applying the operating internal water pressure. There is a large amount of

concrete in the lower part of the horseshoe lining. When the anchors are tensioned, the stress state of this part can be better, which is more conducive to the safety of the structure. When internal water pressure is applied, the ability to resist internal water ultimately comes from the annular anchors. Although the cross-sections of the two linings are different, there is no difference in the layout of the annular anchor, so the ability to resist internal water pressure is basically the same.

Table 5.2 Statistical analysis of stress of horseshoe lining and circular lining

Condition	Position	Calculation results	Horseshoe lining	Circular lining
During anchors tensioning	Overall state	Prestress / MPa	3.0-5.5	3.5-6.0
		Maximum deformation / mm	2.74	2.45
	Near anchorage block-out	Maximum compressive stress /MPa	8.77	9.31
		Maximum tensile stress /MPa	0.95	1.35
During internal water loading	Overall state	Prestress / MPa	1.8-2.5	1.5-2.3
	Near anchorage block-out	Safety status	Yes	Yes

5.5 Discussion and conclusions

After the annular anchors are tensioned, and the prestress of lining along the thickness direction decreases in a gradient from outside to inside. The prestress of the upper part of the lining is evenly distributed, while the local prestress of the lower part will be missing and unevenly distributed due to the influence of the anchorage block-out. With the increase of internal water pressure, the tensile stress zone of the lining arranged in a single row of anchorage block-out extends outward from the anchorage block-out, and the range gradually expands, which is easier to form a through tensile stress zone along the longitudinal direction. The lining with double row anchorage block-outs has better internal pressure resistance.

Along with a practical project, the influence of variable lining thickness on its mechanical characteristics is studied. The MUAA linings generate the circumferential tensile stress by the annular anchors pre-buried inside it, which can resist internal water pressure and prevent it from cracking. The prestress of the lining used to resist internal water pressure is ultimately derived from tensioning the annular anchors; the average thickness of the lining has little effect on its load-bearing capacity. In the elliptical lining, an annulus of a slight tensile stress area appeared on the inner side of the lining, and the crown and invert sections will form the weak area because of the structural bending moments.

The thickness of the horseshoe lining is higher than that of the circular lining at the lower part, so the safety margin of structure is higher, but both can ensure the structural safety. In fact, the influence of cross-section shape on the bearing capacity of the MUAA lining is not obvious. The local stress near the anchorage block-out of horseshoe lining is better, while the overall stress distribution of circular lining is more uniform. Horseshoe lining requires a large increase in the amount of concrete, so the cost performance of circular lining is higher.

References

BEHFARNIA K, BEHRAVAN A, 2014. Application of high-performance polypropylene fibers in concrete lining of water tunnels[J]. *Materials & Design*, 55(3): 274-279.

BIAN K, XIAO M, CHEN J, 2009. Study on coupled seepage and stress fields in the concrete lining of the underground pipe with high water pressure[J]. *Tunnelling and Underground Space Technology*, 24(3): 287-295.

CAO R L, WANG Y J, WANG X G, et al, 2019. Mechanical properties of pre-stressed linings with unbonded annular anchors under high internal water pressure based on large-scale in-situ tests[J]. *Chinese Journal of Geotechnical Engineering*, 8 (8) : 1522-1529.

JIN Q, 2005. *Study on the numerical simulation analysis of prestressed concrete tunnel liner with 3D FEM*[D]. Tianjin: Tianjin university.

JIN Z, 2006. *Design and research on unboned prestressed concrete of pressure tunnels*[D]. Nanjing: Ho Hai university,2006.

KANG J F, SUI C E, WANG X Z, 2014. Structure optimization of the prestressed tunnel lining with unbonded circular anchored tendons[J]. *Journal of Hydraulic Engineering*, 45(1): 103-108.

KORYT'KO N G, LOGINOV P G, MAR'IN A G, et al, 2005. Monolithic Lining for Steel-Pouring Ladles[J]. *Metallurgist*, 49(3-4):91-93.

LEE S Y, LEE S H O, SHIN, D I K, et al, 2007. Development of an inspection system for cracks in a concrete tunnel lining[J]. *Canadian Journal of Civil Engineering*, 34(8): 966-975.

LIN X, SHEN F, 1999. *Technology and Practice Study on prestressed tunnel Lining with unbonded circular anchored tendons of Xiaolangdi project*[M]. Zhengzhou: The yellow river water conservancy press.

MINH N N, CAO P, LIU Z Z, 2021. Contour blasting parameters by using a tunnel blast design mode [J]. *Journal of Central South University*, 28(1): 100-111.

SIMANJUNTAK T D Y F, MARENCE M, MYNETT A E, et al, 2014. Pressure tunnels in non-uniform in situ stress conditions[J]. *Tunnelling and Underground Space Technology*, 42(5): 227-236.

SONG K I, OH T M, CHO G C, 2014. Precutting of tunnel perimeter for reducing blasting-induced vibration and damaged zone - numerical analysis[J]. *KSCE Journal of Civil Engineering*, 18(4):1165-1175.

YAN P, ZHAO Z, LU W, et al, 2015. Mitigation of rock burst events by blasting techniques during deep-tunnel excavation[J]. *Engineering Geology*, 188:126-136.

YAN Z, ZHU F, 2008. Finite element simulation Dongshen water supply improvement project ϕ 4.8 meters large culvert analysis of unbonded prestressed[J]. *Prestress Technology*, 67(2): 3-5.

Chapter 6
Theoretical solution and analysis of combined bearing capacity

6.1 Introduction

Understanding the combined bearing characteristics of lining and surrounding rock have long been a fundamental problem in tunnel and underground engineering (Fulvio, 2010). Tunnel excavation and support during construction are typically considered as rock mass unloading processes (Kimura et al, 1987). In other words, the load (surrounding rock pressure) is transmitted from the surrounding rock to the lining direction, the radial stress of the vertical tunnel wall represents the first principal stress, and the circumferential stress of the parallel tunnel wall represents the third principal stress. If the surrounding rock enters the yield state, the yield type is known as unloading yield. Therefore, the current analysis of the tunnel surrounding rock stability and lining stress is mainly based on the assumption that the surrounding rock enters the unloading yield state (Wu et al, 2011; Bian et al, 2013; Yao et al, 2009).

However, for a tunnel under high internal water pressure during operation, the bearing characteristics of the MUAA lining and surrounding rock are significantly different from those mentioned above (Sui, 2014). When the water pressure in the tunnel is high, the load (internal water pressure) is transferred from the lining to the surrounding rock, and the first principal stress changes from radial stress to circumferential stress. If the surrounding rock enters the yield state, the yield type is expected to be loading yield.

Owing to the different load transfer directions, principal stresses, and plastic states, it is more difficult, as well as important, to understand the joint bearing characteristics of the MUAA lining surrounding rock under internal water pressure.

6.2 Mechanical properties and mathematical modeling

6.2.1 Mechanical properties and mathematical model of surrounding rock

When the mechanical method is used to analyze the deep tunnel engineering, it is usually simplified to solve the plane strain problem (Liu & Du, 2004). As shown in Fig. 6.1, the pressure tunnel has a circular section. If there is initial in-situ stress (q_0) in the surrounding rock at infinity, the load acting on the inner surface of the surrounding rock is p_c. According to lame formula (Yang, 2014) of elasticity theory, the stress of surrounding rock should be:

$$\left.\begin{aligned} \sigma_\theta &= \frac{r_c^2}{r^2} p_c - \left(1+\frac{r_c^2}{r^2}\right) q_0 \\ \sigma_r &= -\frac{r_c^2}{r^2} p_c - \left(1-\frac{r_c^2}{r^2}\right) q_0 \end{aligned}\right\} \tag{6.1}$$

If the surrounding rock of pressure tunnel obeys Mohr-Coulomb elastic-plastic yield criterion (Fig. 6.2), Eq. (6.1) shows that the stress and plastic yield state of surrounding rock are closely related to q_0 and p_c.

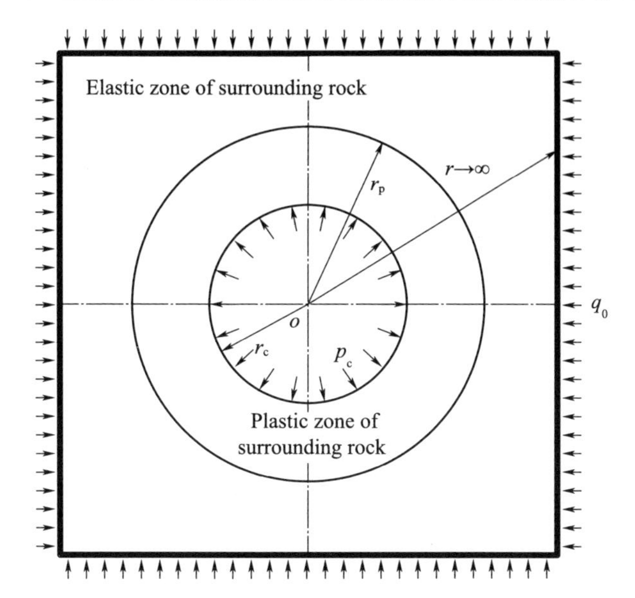

Fig. 6.1 Mechanical mathematical model of pressure tunnel

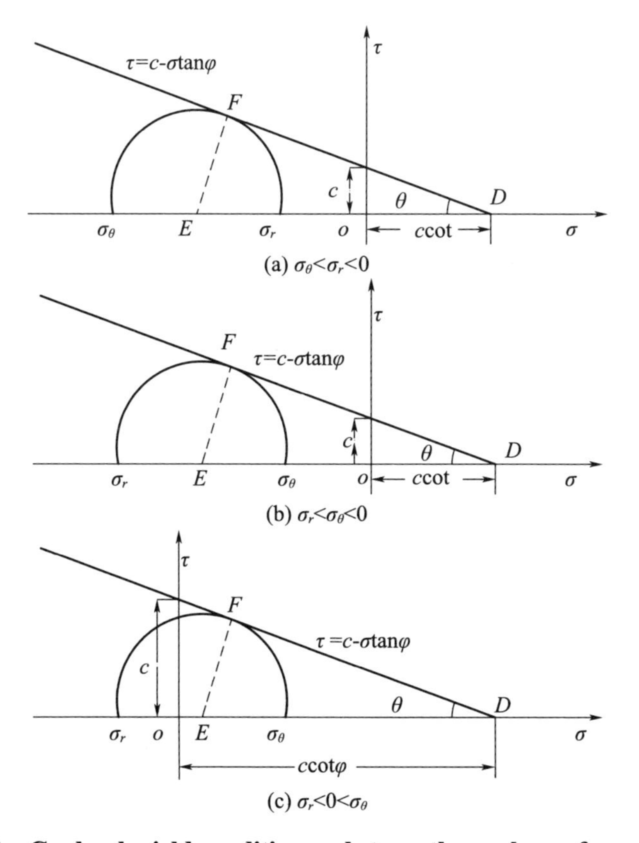

Fig. 6.2 Mohr-Coulomb yield condition and strength envelope of surrounding rock

According to Eq. (6.1), no matter how q_0 and p_c change, there is $\sigma_r < 0$. If $q_0 > p_c$, then $\sigma_\theta < \sigma_r < 0$. Mohr's stress circle is located on the left side of τ axis, and σ_θ is the minimum principal stress. As shown in Fig. 6.2(a), the surrounding rock is in the unloading yield state, and the Mohr Coulomb yield condition is:

$$\frac{\sigma_\theta - \sigma_r}{2} = -\frac{\sigma_\theta + \sigma_r}{2}\sin\varphi + c\cot\varphi \tag{6.2}$$

If $q_0 < p_c < \left(1 + \frac{r^2}{r_0^2}\right) q_0$ then $\sigma_r < \sigma_\theta < 0$. Mohr's stress circle is located on the left side of τ axis, and σ_r is the minimum principal stress. As shown in Fig. 6.2(b), the surrounding rock is in the unloading yield state, and the Mohr Coulomb yield condition is:

$$\frac{\sigma_r - \sigma_\theta}{2} = -\frac{\sigma_\theta + \sigma_r}{2}\sin\varphi + c\cot\varphi \tag{6.3}$$

If $q_0 < \left(1 + \frac{r^2}{r_0^2}\right) q_0 \leqslant p_c$ then $\sigma_r < 0 < \sigma_\theta$. Mohr's stress circle intersects the τ axis, and σ_r is the minimum principal stress. As shown in Fig. 6.2(c), the surrounding rock is in the unloading yield state, and the Mohr Coulomb yield condition is Eq. (6.3) also.

It can be seen from the above that the yield conditions of surrounding rock in Fig. 6.2 (b) and Fig. 6.2 (c) are the same, and the initial in-situ stress is less than the contact stress. Therefore, they can be summarized into the same type. Then, the yield state of surrounding rock can be divided into two cases, namely:

(1) $q_0 > p_c$, σ_θ is the minimum principal stress, and the surrounding rock is in the unloading yield state. This mechanical state is suitable for describing the unloading conditions of tunnel during excavation and pressure tunnel maintenance.

(2) $q_0 < p_c$, σ_r is the minimum principal stress, and the surrounding rock is in the loading yield state. This mechanical state is suitable for describing

the water pressure loading condition in the tunnel during operation.

Therefore, to analyze the plastic yield state of surrounding rock under high internal water pressure, the principal stress sequence and yield state type must be considered into the elastic-plastic analysis, and the yield condition should adopt Eq.(6.3).

6.2.2 Combined bearing characteristics of the MUAA lining and surrounding rock

For the joint bearing system of the MUAA lining and surrounding rock, the internal water pressure load is finally borne by the lining, annular anchor, and surrounding rock (Fig. 6.3), which is manifested in the contact stress between lining and surrounding rock, the equivalent load of annular anchor prestress and internal water pressure. The mechanical properties and directions of these loads are consistent (may be positive or negative). Therefore, the load sharing ratio of surrounding rock (s_r) can be defined as the ratio of contact stress (p_c) to internal water pressure (p_0), that is:

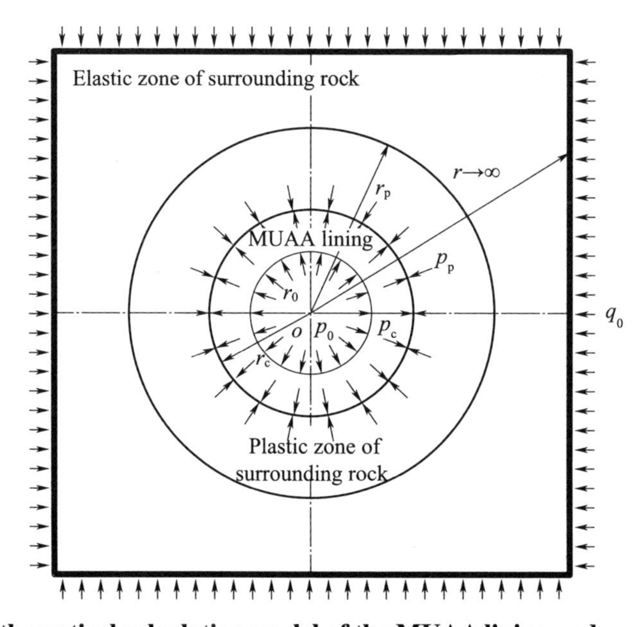

Fig. 6.3 Mathematical calculation model of the MUAA lining and surrounding rock

$$s_r = \frac{p_c}{p_0} \tag{6.4}$$

Therefore, the load sharing ratio of the MUAA lining (s_l) is:

$$s_l = 1 - s_r \tag{6.5}$$

The bearing capacity of the MUAA lining includes annular anchor bearing and concrete bearing. The water load sharing ratio of annular anchor (s_p) is obviously $\frac{p_p}{p_0}$, so the water load sharing ratio (s_c) in concrete should be $1-\frac{p_c}{p_0}-\frac{p_p}{p_0}$.

6.3 Mathematical analysis of combined bearing capacity

6.3.1 Stress and strain of surrounding rock

According to the principle of elasticity (Yang, 2014), the equilibrium differential equation of plane strain problem of circular tunnel is:

$$\frac{\mathrm{d}\sigma_r}{\mathrm{d}r} + \frac{\sigma_r - \sigma_\theta}{r} = 0 \tag{6.6}$$

Where: σ_θ is the circumferential stress; σ_r is radial stress; r is the distance from the point to the origin on the polar coordinates.

The geometric equation is:

$$\varepsilon_\theta = \frac{u}{r};\ \varepsilon_r = \frac{\mathrm{d}u}{\mathrm{d}r} \tag{6.7}$$

Where: ε_θ is circumferential strain; ε_r is radial strain; u is the radial displacement of surrounding rock.

When loading the yield state, σ_r is the first principal stress. Replace the Eq. (6.3) (yield condition) into Eq. (6.6) (the equilibrium of surrounding rock), and obtain the following differential equation according to the boundary conditions on the interface between surrounding rock and lining:

$$\left.\begin{aligned}\frac{\mathrm{d}\sigma_r}{\mathrm{d}r}-\frac{2\sin\varphi\left(c\cot\varphi-\sigma_r\right)}{\left(1+\sin\varphi\right)r}=0\\ \sigma_r\Big|_{r=r_\mathrm{c}}=-p_\mathrm{c}\end{aligned}\right\}\tag{6.8}$$

The variables are separated and integrated, and the Eq. (6.8) is solved to obtain the stress expression of any point in the surrounding rock plastic zone as follows:

$$\left.\begin{aligned}\sigma_\theta=c\cot\varphi-\frac{1-\sin\varphi}{1+\sin\varphi}\left(c\cot\varphi+p_\mathrm{c}\right)\left(\frac{r_\mathrm{c}}{r}\right)^{\frac{2\sin\varphi}{1+\sin\varphi}}\\ \sigma_r=c\cot\varphi-\left(c\cot\varphi+p_\mathrm{c}\right)\left(\frac{r_\mathrm{c}}{r}\right)^{\frac{2\sin\varphi}{1+\sin\varphi}}\end{aligned}\right\}\tag{6.9}$$

According to Eq. (6.1) and Eq. (6.9), at the interface of elastic-plastic zone of surrounding rock ($r = r_\mathrm{p}$), the sum of circumferential stress and radial stress on the side of elastic zone is $-2\sigma_0$ and that on the side of plastic zone is

$$2c\cot\varphi-\frac{2\left(c\cot\varphi+p_\mathrm{c}\right)}{1+\sin\varphi}\left(\frac{r_\mathrm{c}}{r_\mathrm{p}}\right)^{\frac{2\sin\varphi}{1+\sin\varphi}}$$

At the interface of elastic-plastic zone, the stress shall be equal: $\sigma_\theta^e+\sigma_r^e=\sigma_\theta^p+\sigma_r^p$, so it can be obtained that the outer radius (r_p) of surrounding rock plastic zone is:

$$r_\mathrm{p}=r_\mathrm{c}\left[\frac{p_\mathrm{c}+c\cot\varphi}{\left(q_0+c\cot\varphi\right)\left(1+\sin\varphi\right)}\right]^{\frac{1+\sin\varphi}{2\sin\varphi}}\tag{6.10}$$

If $r_\mathrm{c} = r_\mathrm{p}$ is required, the critical contact stress of surrounding rock in loading plastic state is obtained as follows:

$$p_\mathrm{c}^{\mathrm{cri}}=q_0\left(1+\sin\varphi\right)+c\cos\varphi\tag{6.11}$$

According to the theory of elastic-plastic mechanics (Yang, 2014), the elastic strain and plastic strain in the plastic zone of surrounding rock are respectively:

$$
\left.\begin{aligned}
\varepsilon_{\theta}^{\mathrm{e}} &= \frac{1}{E_{\mathrm{d}}}\left[\left(1-\frac{1}{2}\mu_{\mathrm{d}}\right)\sigma_{\theta}-\frac{3}{2}\mu_{\mathrm{d}}\sigma_{r}\right] \\
\varepsilon_{r}^{\mathrm{e}} &= \frac{1}{E_{\mathrm{d}}}\left[\left(1-\frac{1}{2}\mu_{\mathrm{d}}\right)\sigma_{r}-\frac{3}{2}\mu_{\mathrm{d}}\sigma_{\theta}\right] \\
\varepsilon_{\theta}^{\mathrm{p}} &= \frac{\psi}{4G_{\mathrm{d}}}\left(\sigma_{\theta}-\sigma_{r}\right) \\
\varepsilon_{r}^{\mathrm{p}} &= \frac{\psi}{4G_{\mathrm{d}}}\left(\sigma_{r}-\sigma_{\theta}\right)
\end{aligned}\right\} \tag{6.12}
$$

Where: $\varepsilon_{\theta}^{\mathrm{e}}$ is the circumferential elastic strain and $\varepsilon_{r}^{\mathrm{e}}$ is the radial elastic strain; $\varepsilon_{\theta}^{\mathrm{p}}$ is the circumferential plastic strain and $\varepsilon_{r}^{\mathrm{p}}$ is the radial plastic strain; E_{d} is the elastic modulus of surrounding rock; μ_{d} is the Poisson's ratio of surrounding rock; ψ is the plastic potential function; G_{d} is the shear modulus of surrounding rock.

The strain of surrounding rock in plastic zone is the sum of elastic strain and plastic strain, i.e.:

$$
\begin{Bmatrix}\varepsilon_{\theta} \\ \varepsilon_{r}\end{Bmatrix}=\begin{Bmatrix}\varepsilon_{\theta}^{\mathrm{e}} \\ \varepsilon_{r}^{\mathrm{e}}\end{Bmatrix}+\begin{Bmatrix}\varepsilon_{\theta}^{\mathrm{p}} \\ \varepsilon_{r}^{\mathrm{p}}\end{Bmatrix} \tag{6.13}
$$

Then, combining Eq. (6.7) and Eq. (6.13), the geometric equation of surrounding rock deformation in plastic zone can be obtained:

$$
\left.\begin{aligned}
\frac{\mathrm{d}u}{\mathrm{d}r} &= \frac{1}{E_{\mathrm{d}}}\left[\left(1-\frac{1}{2}\mu_{\mathrm{d}}\right)\sigma_{r}-\frac{3}{2}\mu_{\mathrm{d}}\sigma_{\theta}\right]-\frac{\psi}{4G_{\mathrm{d}}}\left(\sigma_{\theta}-\sigma_{r}\right) \\
\frac{u}{r} &= \frac{1}{E_{\mathrm{d}}}\left[\left(1-\frac{1}{2}\mu_{\mathrm{d}}\right)\sigma_{\theta}-\frac{3}{2}\mu_{\mathrm{d}}\sigma_{r}\right]+\frac{\psi}{4G_{\mathrm{d}}}\left(\sigma_{\theta}-\sigma_{r}\right)
\end{aligned}\right\} \tag{6.14}
$$

Substituting Eq. (6.9) into Eq. (6.14), the following differential equation can be obtained:

$$
\frac{\mathrm{d}u}{\mathrm{d}r}+\frac{u}{r}=\frac{4\mu_{\mathrm{d}}-2}{E_{\mathrm{d}}}\left[\frac{p_{\mathrm{c}}+c\cot\varphi}{1+\sin\varphi}\left(\frac{r_{c}}{r}\right)^{\frac{2\sin\varphi}{1+\sin\varphi}}-c\cot\varphi\right] \tag{6.15}
$$

According to the second formula in Eq. (6.9) and Eq. (6.14), the radial deformation on the boundary interface ($r = r_p$) of the elastic-plastic zone can be obtained as follows:

$$\left.\frac{u}{r}\right|_{r=r_\mathrm{p}} = -\frac{(1-2\mu_\mathrm{d}-\sin\varphi-\mu_\mathrm{d}\sin\varphi)(p_\mathrm{c}+c\cot\varphi)}{E_\mathrm{d}(1+\sin\varphi)}\left(\frac{r_\mathrm{c}}{r_\mathrm{p}}\right)^{\frac{2\sin\varphi}{1+\sin\varphi}}+\frac{(1-2\mu_\mathrm{d})c\cot\varphi}{E_\mathrm{d}} \tag{6.16}$$

Eq. (6.16) is the definite solution condition of Eq. (6.15), and the solution of Eq. (6.15) can be obtained:

$$\begin{aligned} u = & \frac{r(1-2\mu_\mathrm{d})}{E_\mathrm{d}}\left[c\cot\varphi-(p_\mathrm{c}+c\cot\varphi)\left(\frac{r_\mathrm{c}}{r_\mathrm{p}}\right)^{\frac{2\sin\varphi}{1+\sin\varphi}}\right] \\ & +\frac{(2-\mu_\mathrm{d})(p_\mathrm{c}+c\cot\varphi)r_\mathrm{p}^2\sin\varphi}{E_\mathrm{d}(1+\sin\varphi)r}\left(\frac{r_\mathrm{c}}{r_\mathrm{p}}\right)^{\frac{2\sin\varphi}{1+\sin\varphi}} \end{aligned} \tag{6.17}$$

Then, according to Eq. (6.17), the radial displacement on the interface between surrounding rock and lining ($r = r_c$) is:

$$u_{r_\mathrm{c}}^1 = \frac{r_\mathrm{c}(2-\mu_\mathrm{d})\sin\varphi(q_0+c\cot\varphi)^{-\frac{1}{\sin\varphi}}}{E_\mathrm{d}}\left(\frac{p_\mathrm{c}+c\cot\varphi}{1+\sin\varphi}\right)^{\frac{1+\sin\varphi}{\sin\varphi}}-\frac{r_\mathrm{c}(1-2\mu_\mathrm{d})p_\mathrm{c}}{E_\mathrm{d}} \tag{6.18}$$

6.3.2 Stress and strain of the MUAA lining

The mechanical characteristics of the MUAA lining conform to the thick wall cylinder theory of elasticity (Liu & Du, 2004). Therefore, according to the principle of elasticity, the circumferential stress and radial stress of lining are respectively:

$$\left.\begin{matrix}\sigma_\theta \\ \sigma_r\end{matrix}\right\} = \frac{p_0 r_0^2-(p_\mathrm{c}+p_\mathrm{p})r_\mathrm{c}^2}{r_\mathrm{c}^2-r_0^2} \mp \frac{r_0^2 r_\mathrm{c}^2(p_\mathrm{c}+p_\mathrm{p}-p_0)}{(r_\mathrm{c}^2-r_0^2)r^2} \tag{6.19}$$

The radial displacement of lining is:

$$u=\frac{\left(1+\mu_{c}\right)\left(1-2\mu_{c}\right)\left(p_{0}r_{0}^{2}-p_{c}r_{c}^{2}-p_{p}r_{c}^{2}\right)r}{E_{c}\left(r_{c}^{2}-r_{0}^{2}\right)}-\frac{r_{0}^{2}r_{c}^{2}\left(1+\mu_{c}\right)\left(p_{c}+p_{p}-p_{0}\right)}{E_{c}\left(r_{c}^{2}-r_{0}^{2}\right)r} \tag{6.20}$$

Therefore, according to Eq. (6.20), under the combined action of prestress (p_p) and internal water pressure (p_0), the radial displacement of ($r = r_c$) at the interface between surrounding rock and lining is:

$$u_{r_c}^{2}=\frac{\left(1+\mu_{c}\right)\left[2\left(1-\mu_{c}\right)r_{0}^{2}p_{0}-\left(r_{c}^{2}-2\mu_{c}r_{c}^{2}+r_{0}^{2}\right)\left(p_{c}+p_{p}\right)\right]r_{c}}{E_{c}\left(r_{c}^{2}-r_{0}^{2}\right)} \tag{6.21}$$

6.3.3 Displacement of the MUAA lining

The application of prestress will lead to the overall shrinkage of the MUAA lining and the separation of the top from the surrounding rock. It is necessary to fill the gap through contact grouting to ensure the joint bearing of lining and surrounding rock. The contact grouting effect will greatly affect the joint stress of lining and surrounding rock. The filling coefficient (λ) is defined to represent the grouting effect. When $\lambda = 0$, there is no contact grouting and the surrounding rock is not stressed. When $\lambda > 0$, there is contact grouting and the lining and surrounding rock are jointly stressed. When $\lambda = 1$, the effect of contact grouting is ideal, and the combined stress state of lining and surrounding rock is good. After contact grouting, when calculating the final radial displacement of the MUAA lining, Eq. (6.21) shall use $(1-\lambda)p_p$ instead of p_p.

According to the displacement coordination conditions of lining and surrounding rock at the contact surface, the following can be obtained:

$$u_{r_c}^{1}=u_{r_c}^{2} \tag{6.22}$$

Substituting Eq. (6.18) and Eq. (6.21) into Eq. (6.22) and using $(1-\lambda)p_p$ instead of p_p, the contact stress (p_c) between lining and surrounding rock can be obtained, which meets the Eq. (6.23):

$$E_c r_0\left(r_c^2-r_0^2\right)\left[(2-\mu_d)\sin\varphi(q_0+c\cot\varphi)^{-\frac{1}{\sin\varphi}}\left(\frac{p_c+c\cot\varphi}{1+\sin\varphi}\right)^{\frac{1+\sin\varphi}{\sin\varphi}}-(1-2\mu_d)p_c\right]-$$

$$E_d r_c(1+\mu_c)\left[2(1-\mu_c)r_0^2 p_0-\left(r_c^2-2\mu_c r_c^2+r_0^2\right)\left(p_c+p_p-\lambda p_p\right)\right]=0$$

(6.23)

It should be noted that p_c is the implicit solution in Eq. (6.23). However, for tunnel engineering, the relevant parameters in the formula are known, so it is not difficult to solve in practical application. After the contact stress (p_p) is obtained, the load sharing ratio of surrounding rock and the MUAA lining under combined bearing conditions can be calculated according to Eq. (6.4) and Eq. (6.5) respectively.

6.3.4 Solving process of mathematical model

The mechanical properties of the MUAA lining and surrounding rock under combined bearing conditions mainly include:

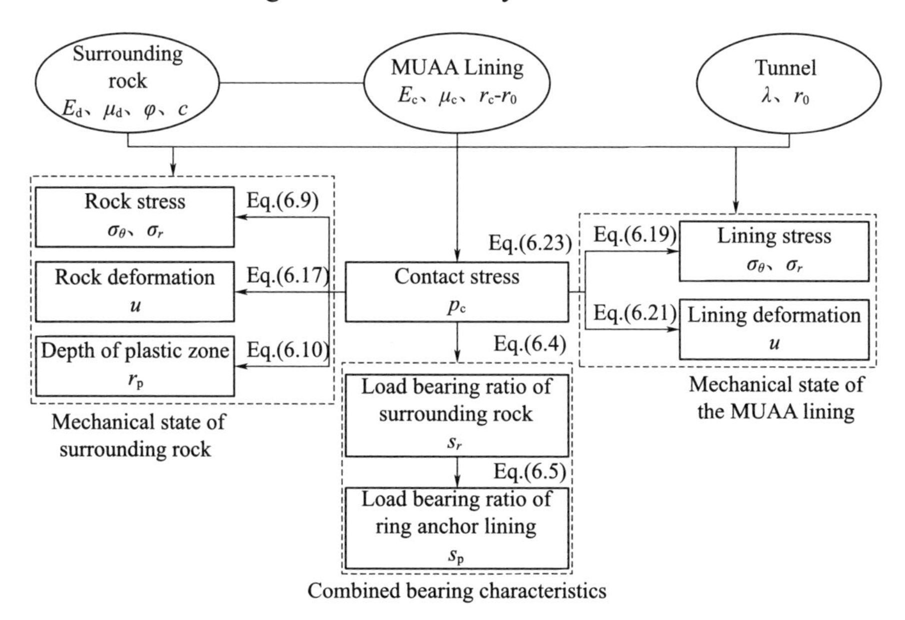

Fig. 6.4 Solving process of mathematical model

(1) Stress, deformation, and development depth of plastic zone of surrounding rock.

(2) Stress and deformation of lining.

(3) Share proportion of internal water load between lining and surrounding rock.

The mathematical models established in this study can solve the above variables. The specific mathematical model parameter input and solution process are shown in Fig. 6.4.

6.4 Example and verification

6.4.1 Engineering case

The engineering case is the MUAA lining at Songhua River Water Diversion Works, which is a large-scale long-distance water transfer project for solving the water supply problem in north-eastern China (Fig. 3.2 and Fig. 6.5). The parameters of pressure tunnel, surrounding rock and MUAA lining are listed in Table 2.1.

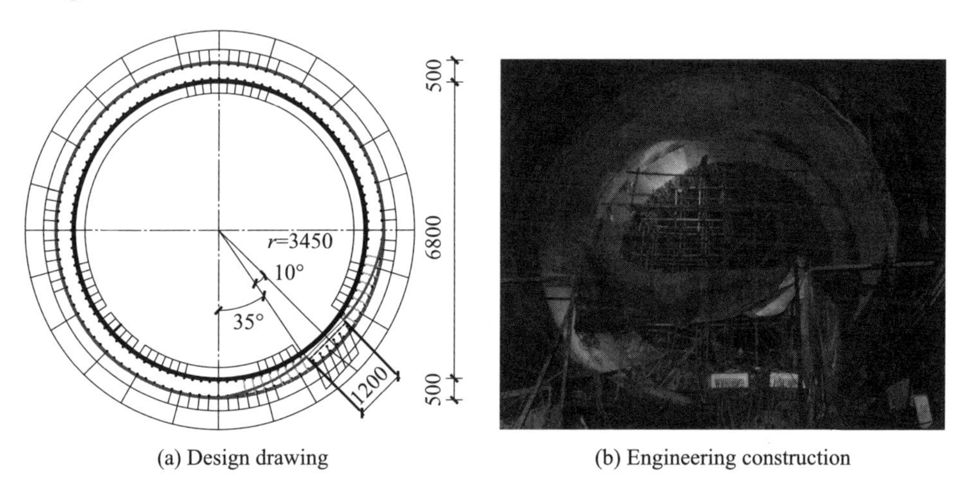

(a) Design drawing (b) Engineering construction

Fig. 6.5 Structural and construction drawings of the MUAA lining (unit: mm)

6.4.2 Example analysis

The mathematical model of combined bearing capacity of the MUAA lining and surrounding rock is used for case analysis. The relationship between circumferential stress (σ_θ) and radial stress (σ_r) of surrounding rock within different contact stress conditions and depth is shown in Fig. 6.6. The calculation results show that in any case, there are $\sigma_\theta < \sigma_r$. Therefore, it is reasonable to assume the radial stress of surrounding rock as the first principal stress in the mathematical model.

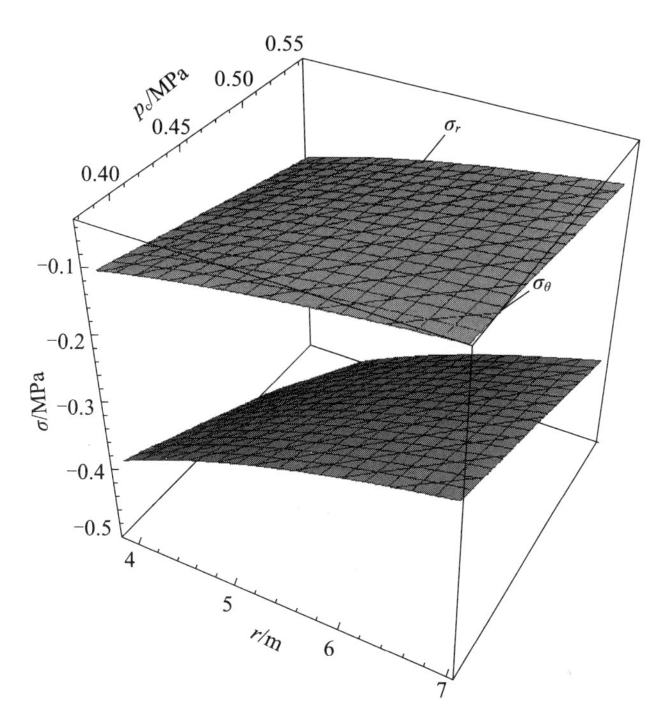

Fig. 6.6 Circumferential stress and radial stress of surrounding rock

Fig. 6.7 is the variation curves of contact stress (p_c) and surrounding rock load sharing coefficient (s_r) with internal water pressure (p_0). With the p_0 increase, the bearing capacity of surrounding rock increases gradually, while s_r decreases gradually. As shown in Fig. 6.8, when the lining is in ideal contact with the surrounding rock ($\lambda = 1$), the internal water pressure increases from 1.0 MPa to 2.0 MPa and 3.0 MPa, while p_c will increase from 0.38 MPa to 0.48 MPa and 0.55 MPa. At the same time, s_r decreases from 0.38 to 0.24 and 0.18 respectively. The bearing capacity of the MUAA lining depends on the concrete and prestressed annular anchor. No matter how the internal water pressure changes, the load proportion of concrete remains unchanged, and the coefficient is 0.11.

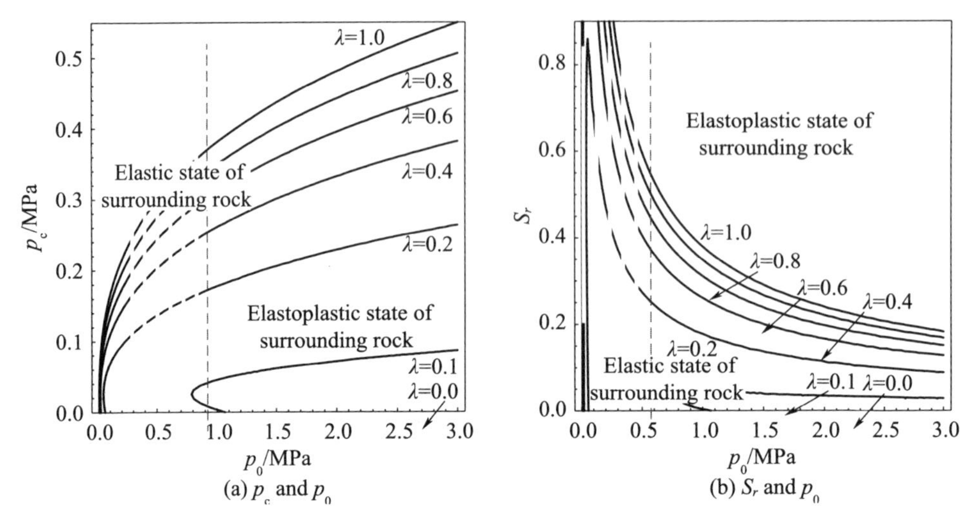

Fig.6.7 The variation curves of p_c and s_r with p_0

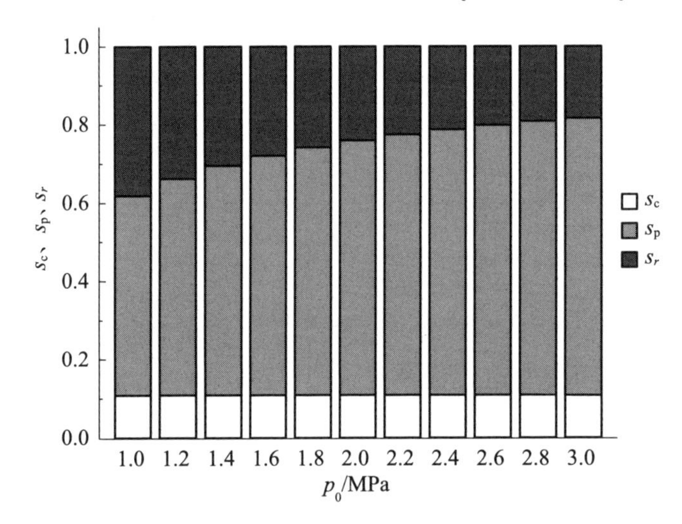

Fig.6.8 Load sharing ratios under different internal water pressure

The load sharing ratio between the annular anchor and surrounding rock will change one after another. When the internal water pressure is low, the load sharing ratio of surrounding rock is very high. The self-bearing function of surrounding rock should be used as much as possible in the design of pressure tunnel. When the internal water pressure is high, the bearing capacity of surrounding rock is limited, and the prestress of annular anchor should be increased. Therefore, it is obviously inappropriate to ignore the bearing effect of surrounding rock. If the internal water pressure load is completely borne by the annular anchor, the design value of prestress will be too large. This not only wastes resources, but also has large prestress value and unfavorable mechanical state of lining. For example, under the action of internal water pressure and prestress, the best stress state of lining should be that the circumferential stress of lining is zero. When the internal water pressure is 1.0 MPa, the design value of corresponding prestress of annular anchor can be 0.62 MPa.

6.5 Analysis of combined bearing characteristics

In order to better understand the combined bearing characteristics of the MUAA lining and surrounding rock, the parameter sensitivity analysis of the new mathematical model is carried out. During the calculation, the internal water pressure and prestress are set to 1.0 MPa, and the parameters of surrounding rock, lining and tunnel are considered (Table 6.1). The variation curves of surrounding rock load sharing coefficient (s_r) are shown in Fig. 6.9 and Fig. 6.10. The calculation results show that:

(1) The load sharing is mainly affected by elastic modulus (E_d) and Poisson's ratio (m_d) of surrounding rock. s_r exceeds 0.4 for class I rock mass ($E_d > 20$ GPa; $m_d < 0.2$) and class II rock mass (10 GPa $< E_d <$ 20 GPa; $0.2 < m_d < 0.25$).

(2) The changes of elastic modulus (E_c) and Poisson's ratio (μ_c) of lining have little effect on s_r, and the fluctuation of s_r does not exceed 0.04.

(3) With the increase of lining thickness (r_c-r_0), s_r gradually decreases. However, when the lining thickness increases to 0.5 m, s_r basically converges.

(4) The influence of tunnel radius (r_0) on s_r is also obvious. When r_0 increases from 1.0 m to 5.0 m, s_r increases from 0.24 to 0.32, an increase of 25%.

Table 6.1 Calculation parameters of mathematical model

Factors	Surrounding rock		Lining			Tunnel	
	Elastic modulus	Poisson's ratio	Elastic modulus	Poisson's ratio	Thickness	Filling coefficient	Radius
Symbol / unit	E_d/GPa	μ_d	E_c/GPa	μ_c	r_c-r_0/m	λ	r_0/m
Minimum	0.3	0.38	31.7	0.175	0.2	0.2	1.0
Median	6	0.30	36.0	0.190	0.7	0.6	3.0
Maximum	33	0.20	39.3	0.205	1.2	1.0	5.0

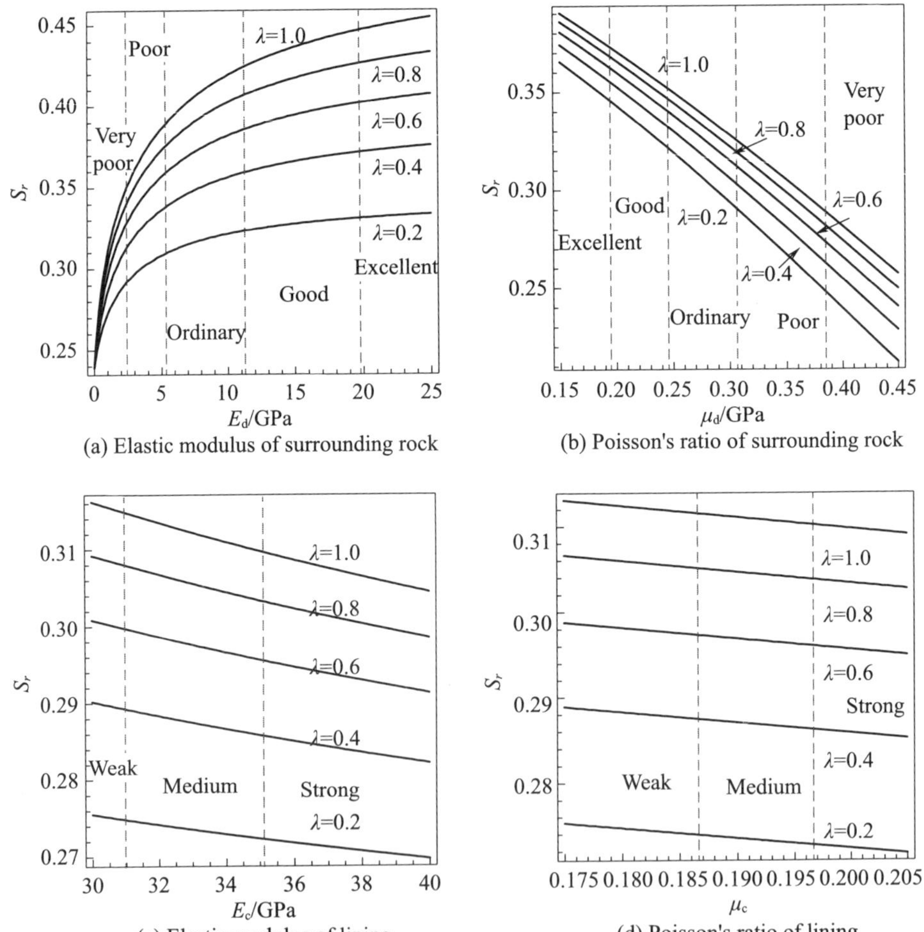

(a) Elastic modulus of surrounding rock

(b) Poisson's ratio of surrounding rock

(c) Elastic modulus of lining

(d) Poisson's ratio of lining

Fig.6.9(1) The variation curves of surrounding rock load sharing coefficient

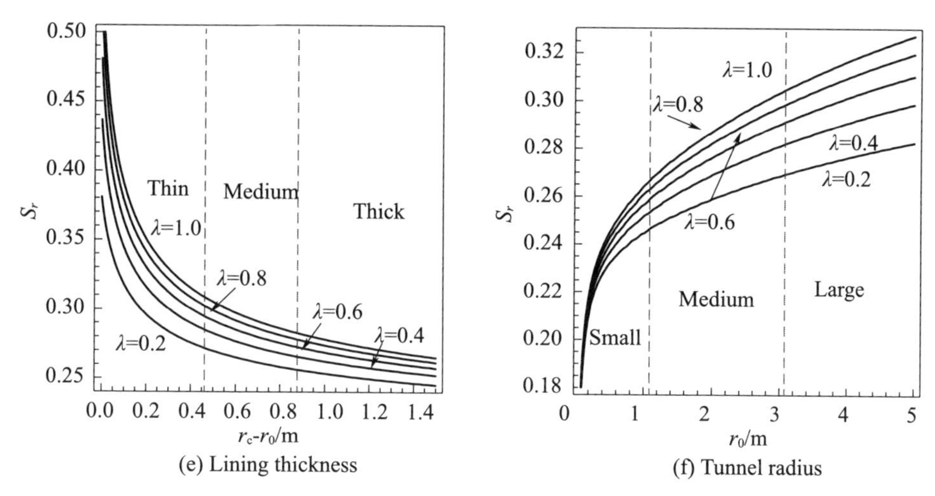

(e) Lining thickness (f) Tunnel radius

Fig.6.9(2) The variation curves of surrounding rock load sharing coefficient

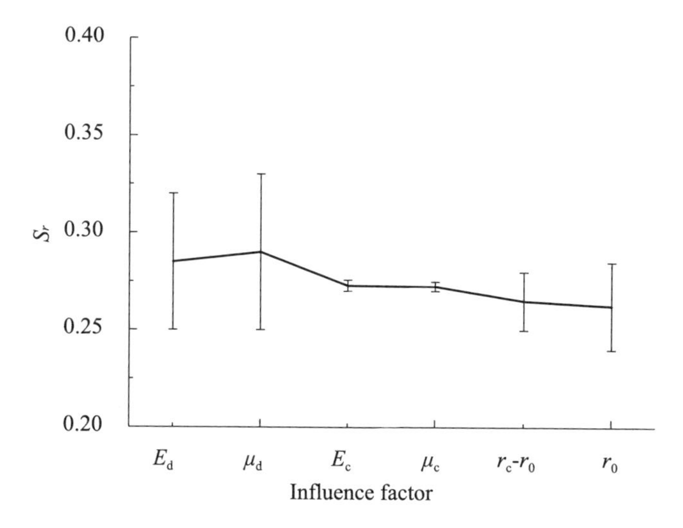

Fig.6.10 Effects of various factors on surrounding rock load sharing coefficient

In addition, when the lining is in ideal contact with the surrounding rock ($\lambda = 1$), s_r exceeds 0.25. In other words, the internal water pressure load borne by surrounding rock exceeds 25%. It is inappropriate to completely ignore the load sharing of surrounding rock in the design of existing projects. Therefore, in the design of the MUAA lining, it is not necessary to require that the internal water pressure is equal to the equivalent prestress of annular anchor (p_p), and the p_p value can be appropriately reduced.

6.6 Discussion and conclusions

The combined bearing characteristics of the MUAA lining and surrounding rock are related to many factors, such as prestress, internal and external water pressure, surrounding rock pressure, tunnel radius, lining thickness, and elastic modulus. Field tests or model test methods are often considered to be more similar to actual engineering, but consume a significant amount of manpower and material resources. Numerical tests demonstrate repeatability and visualization and are widely used in the study of the mechanical properties of the lining. A variety of orthogonal tests is required to study the relationship among multiple factors. By establishing an accurate functional relationship among various factors through an analytical method, the problem can be solved quickly, with universal applicability for engineering design.

The interaction characteristics between the prestressed MUAA lining and the surrounding rock are analyzed. The phenomenon, causes, and judgment criteria of the sequential transformation of the yield state and principal stress of the surrounding rock under high internal water pressure are proposed. By using the radial equivalent load expression of the annular anchor tension, the load types of the prestresses, internal water pressure, and surrounding rock pressure can be unified. Thus, a mathematical model of the combined bearing characteristics of the MUAA lining and surrounding rock is established to realize a unified analysis of the stress and strain.

The load-sharing coefficient of the surrounding rock is defined and analyzed. The main factors affecting the load bearing of the surrounding rock are the elastic modulus, Poisson's ratio, and tunnel diameter. If the backfill grouting after the prestressed ring anchor lining is significantly dense, the surrounding rock can share the internal water pressure load. It is inappropriate to completely ignore the bearing capacity of the surrounding rock in the design. Properly reducing the prestress of the MUAA lining and using the bearing capacity of the surrounding rock are expected to increase the structural safety.

References

BIAN Y W, XIA C C, XIAO W M, et al, 2013. Visco-elastoplastic solutions for circular tunnel considering stress release and softening behaviour of rocks [J]. *Rock and Soil Mechanics*, 341(1):211-221.

FAHIMIFAR A, SOROUSH H, 2005. A theoretical approach for analysis of the interaction between grouted rockbolts and rock masses[J]. *Tunneling and Underground Space Technology*, 20: 333–343.

FULVIO T, 2010. Sequential excavation, NATM and ADECO: what they have incommon and how they differ[J]. *Tunnelling and underground Space Technology*, 25(3): 145–265.

Kimura F, Okabayshi N, Kawamoto T, 1987. Tunnelling Through Squeezing Rock in Two Large Fault Zones of the Enasan Tunnel II[J]. *Rock Mechanics and Rock Engineering*, 20(3):151-166.

LIU B G, DU X D, 2004. Visco elastic analysis on iteraction between supporting structure and surrounding rocks of circular tunnel[J]. *Chinese Journal of Rock Mechanics and Engineering*, 23(4): 561-564.

SIMANJUNTAK T D Y F, MARENCE M, MYNETT A E, et al, 2014. Pressure tunnels in non-uniform in-situ stress conditions[J]. *Tunnelling & Underground Space Technology*, 42(5): 227–236.

SUI C E, 2014. *Stress analysis and safety evaluation of Xiaolangdi pre-stressed tunnel lining with unbounded circular anchored tendons*[D]. Tianjin: Tianjin University.

WU S C, PAN D G, GAO Y T, 2011. Analytic solution for rock-liner interaction of deep circular tunnel[J]. *Engineering Mechanic*, 28(3): 136–142.

YANG B Y, 2014. *Engineering elastic-plastic mechanics*[M]. Beijing: China Machine

Press.

YAO G S, LI J P, GU S C, 2019. Analytic solution to deformation of soft rock tunnel considering dilatancy and plastic softening of rock mass [J]. *Rock and Soil Mechanics*, 30(2): 463-467.